THE REMINISCENCES OF
JOHN B. JERVIS

ENGINEER OF THE OLD CROTON

A YORK STATE BOOK

THE REMINISCENCES OF
JOHN B. JERVIS

ENGINEER OF THE OLD CROTON

Edited, with Introduction, by
NEAL FITZSIMONS

Foreword by ROBERT VOGEL

SYRACUSE UNIVERSITY PRESS

CONTENTS

ILLUSTRATIONS

FOREWORD

THE CONTRIBUTIONS of John B. Jervis to the growth of the United States were great in scope and many in number. But since civil engineers are often neglected by posterity, Jervis' name and his contributions to history drifted into obscurity even before his death in 1885.

The exploits of soldiers, the tribulations of politicians, the sagas of explorers—these are well known and admired by all, but the achievements of engineers are often taken for granted. Happily, there appears to be an increasing awareness by historians of the impact of engineers on our society, and one hopes that future history books will reflect this. The present volume is a welcome addition to the limited number of biographical works about American engineers. Jervis is of especial importance in American engineering history for several reasons. First, he was among the leading figures of mid-nineteenth-century engineering in the United States. Second, he represents that breed of engineer who, with almost no formal education, rose to professional leadership through practical experience and assiduous independent study. Third, Jervis, though styling himself a civil engineer, was by present definitions both a civil and a mechanical engineer. Further, he engaged in a wide range of projects—canals, railroads, and waterworks—in contrast to practitioners of today who are far more highly specialized.

Aside from his technical achievements, Jervis is interesting as a man of his times. His ideals of honesty, integrity, and frugality were clear cut, as were his religious convictions and his political opinions. How would such a personality fare in contemporary America? Each reader must speculate for himself. Although this book emphasizes the engineering side of his life, there is just enough comment of a personal nature to induce this type of speculation.

The importance of this book, however, lies in its details of the development of American civil engineering during its period of most rapid growth. From this volume, the scholar can gain direct insight into this development, and the general reader can enjoy the fascinating story of how Jervis' great engineering works were really built.

Washington, D.C.
Fall 1970

ROBERT VOGEL, *Curator, Division of*
Mechanical & Civil Engineering
Smithsonian Institution

PREFACE

THIS BOOK is based on a series of autobiographical sketches, entitled " 'Facts and Circumstances in the Life of John B. Jervis,' by himself," probably begun after the author's fiftieth year and continued past his eightieth year. While doing research at the Smithsonian Institution on the life of Benjamin Wright, the Father of American Civil Engineering, I found there a copy of Jervis' original manuscript held by the Jervis Library in Rome, New York, where Jervis' thousands of letters and papers are housed. The Jervis papers have been indexed by the Jervis Library and are on microfilm.

With the encouragement of William A. Dillon, Director, the Trustees of the Jervis Library, and the C. C. Reid Memorial Fund, whose generous assistance has facilitated publication, I gathered background material for an Introduction and provided footnotes to explain technical terms to lay readers and to spare them unnecessary diversions to other reference works.

My Introduction is intended to give a broad background to Jervis' text and to acquaint the reader with the technical significance as well as the present status of his works. For convenience, the narrative of the Introduction generally follows the sequence of the subsequent chapters; hence the topics in the Introduction correspond to main chapter titles.

Working from Samuel Hyde Beach's transcript of Jervis' memoirs, along with the original written by Jervis himself—both of which are located in the Jervis Library—I rearranged the parts slightly, combining chapters to obtain more continuity, dividing the book into thirteen chapters instead of the original twenty-four sections. I have deleted some repetitious phrases and modernized the spelling and punctuation. The last two chapters, "Later Projects" and "Professionalism and the Engineer," appear in the Beach version but not in Jervis' original text. The Beach family of Rome, New York, is related to the Jervis family through the marriage of John Bloomfield Jervis' sister Susannah Maria to Samuel Beach, M.D., in 1817. It was through the family of Bloomfield Jervis Beach, their oldest son, that the transcript of Jervis' memoirs was handed down to the present via Samuel Hyde Beach.

Jervis' account of his experiences, his commentaries on several important engineering controversies, and his careful descriptions of engineering techniques should be of interest not only to historians of technology and Jervis' professional legatees in engineering, but also to others who wonder about how our great public works—railroads, canals, dams, and waterworks—were constructed, and to those who want to gain a greater appreciation for the men who literally built our country—men like the distinguished American civil engineer, John Bloomfield Jervis.

Kensington, Maryland   NEAL FITZSIMONS, *Fellow*
Fall 1970   *American Society of Civil Engineers*

INTRODUCTION

THE BOUNDARIES of New York, New Jersey, and Pennsylvania meet at a point on the banks of the Delaware River. Here, overlooking the river's ephemeral rapids, is the unpretentious city of Port Jervis, New York. Among its landmarks is a small fieldstone building. Nearby, a roadside marker tells the curious that this obscure old structure is:

DECKER FORT
Home of Martinus Decker
Burned by Brant's Raiders
July 20, 1779. Rebuilt
1793. Headquarters of
John B. Jervis, 1826–28

It is indeed ironic that the man after whom the city itself is named receives such inadequate recognition. But perhaps this is not unusual, for there are very few pages of history written about men like Jervis, civil engineers who conceived, planned, and directed the construction of the great public works of the United States.

The period of 1826–28 fortuitously mentioned on the historical marker were the critical years of Jervis' professional life, because it was then, as chief engineer for the Delaware and Hudson Canal project, he had his first opportunity to prove he could manage a major engineering enterprise successfully. With his professional reputation enhanced by this achievement, Jervis went on to become Americas leading consulting engineer of the antebellum Era.

John B. Jervis descended from old pioneer stock which crossed Long Island Bay from Connecticut to Huntington, New York, about 1653. Shortly after Washington's defeat on Long Island in August, 1776, the British began their seven-year occupation of Huntington. Jervis' grandfather, Jonathan Jarvis, took a loyalty oath to the king in 1778 and remained in the town during and following the war. His father, Timothy Jarvis, remained on the family farm until he was about thirty-one years old when he sold it and moved to Fort Stanwix, New York, with his wife Phebe and their baby son, John. This was in 1797. The house where Jervis was born still stands on the southeast corner of Park Avenue and Woodhull Street in Huntington, New York. Records indicate it was built around 1688 by William Jarvis, a whaling captain and distant relative of Jervis.

It appears that the entire Jarvis family changed their name to Jervis sometime before 1820. Family tradition says it was John who instigated the change. One could speculate that it could have been a patriotic gesture made during the War of 1812 in reaction to the loyalty oath signed by his grandfather during the Revolution.

John Jervis' formal education prepared him for little more than the life of a small farmer or carpenter like his father. But a combination of chance and his alert mind prevented his talents from remaining untapped. Among his father's acquaintances and fellow townsmen was Benjamin Wright, now recognized as the Father of American Civil Engineering. It was by chance that young Jervis was hired by Wright as a temporary field hand to clear brush for the surveyors of the Erie Canal.

Jervis was raised to the rank of resident engineer by Wright in 1819. His assignment to take over David S. Bates's division showed Wright's early confidence in Jervis. His new duties included the supervision of minor construction and necessitated learning the elements of mechanics and hydraulics.

Jervis often mentions John Bloomfield, his uncle and namesake. Dozens of letters in the 1820's, 30's, and 40's testify to Jervis' devotion to his uncle whom he seemed to have held in the greatest admiration. The uncle's possession of the *Edinburgh En-*

cyclopaedia was indeed fortunate for Jervis, since it was an excellent reference on engineering subjects.

Uncle John also seems to have been influential in shaping Jervis' religious views, encouraging him to maintain a strong Calvinistic orientation.

Jervis' career accelerated about 1820. At that time, he became an established member of the Erie Canal's engineering staff, rapidly gaining the confidence of his superiors, especially Wright and Canvass White. Contacts with some of New York State's influential citizens became more frequent after 1820. Among these men was William C. Bouck, who later became governor of New York (1842–44) and with whom Jervis carried on an extensive correspondence in later years. Jervis had very strong opinions on such matters as responsibility of men in public or corporate employ. They are found in his book *Labor and Capital* (1877) and in some of his other writings. He tended to extol the Franklinian virtues of frugality and temperance; Jervis' pride in keeping his accounts was typical of Franklin's "virtue" of order.

Benjamin Wright was apparently a good judge of character, for he had selected and developed a staff of engineers on the Erie Canal who, as individuals, were to become leaders on many major antebellum engineering works. Among these leaders were Canvass White, who planned canals in Connecticut, New Jersey, and Pennsylvania; Nathan S. Roberts, who built canals in Delaware, Maryland, and Pennsylvania; and David S. Bates, who built canals in Delaware, Maryland, Pennsylvania, and Ohio.

Jervis worked not only under Wright, but also White, Roberts, and Bates, and he became the most famous "graduate" of the Erie "School of Engineering." His first post-graduation employment was the Delaware and Hudson project, and before he left it, his reputation as a brilliant engineer was to rival that of his old mentor, Wright.

The Delaware and Hudson Company was the brain-child of the Wurtz brothers, Maurice and William, whose extensive land holdings in northeastern Pennsylvania had large anthracite deposits. The Wurtz brothers were anxious to develop a market in their hometown of Philadelphia and also in the rapidly growing Hudson Valley, especially New York. Land transport of bulk items, like coal, precluded major long-range markets, though railroads a decade later were to change this. The water transport route in the early 1820's required a journey from the mines of about two hundred miles down the Delaware River, sixty miles through the Delaware Bay and one hundred and fifty miles up the Atlantic Coast to New York, or over four hundred miles. Transshipment in different kinds of boats further complicated the problem.

In 1823, Benjamin Wright was engaged to study the problem of water transportation, and his recommendation, and "independent" canal system—that is, a system which did not use the river directly—was selected. It was to go from Kingston, New York, on the Hudson River, fifty-nine miles southwest to the Delaware River near Carpenter's Point, then forty-six miles northeast along the Delaware River to what is now Honesdale, Pennsylvania. Later, Carpenter's Point was renamed Port Jervis. Honesdale was named after Phillip Hone, the first president of the company and mayor of New York. The use of anthracite (versus bituminous) coal as a locomotive fuel was an issue Jervis had to face in the design of early locomotives. Undoubtedly, his association with the Delaware and Hudson Company influenced his decisions.

Not mentioned by Jervis was another great American engineer, John Augustus Roebling, who was called upon by the D&H Company to supply his newly invented (1841) wire rope for the gravity railroad at Honesdale in May of 1841. This rope proved so reliable that Roebling was asked to build four large aqueducts supported by his wire rope during the period of canal rebuilding and enlarging from 1842 to 1851 under the supervision of Jervis' protégé, James Archbald. Jervis himself designed one major structure for the original canal—a two-span 224-ft. aqueduct supported by a wood truss over the Neversink River at Cuddebackville, New York. This structure was replaced by Roebling in 1851, after being in service around twenty-five years. The canal system itself was closed down in 1898.

On October 11, 1969, the abandoned D&H Canal was declared

a national historic site by the National Park Service. Much of the route is still clearly visible,[1] and the building which Jervis used for his headquarters, a simple stone house known as Decker Fort, remains standing in Port Jervis.

Although principally a civil engineer, Jervis' versatility was shown by his brilliance in what would now be termed mechanical engineering—in particular, his contributions to the development of locomotives.

Early attempts at developing a steam locomotive for railway purposes began to be successful about 1804 when the Cornish Giant, Richard Trevithick, built his "Gateshead" locomotive. A decade later George Stephenson's "Blucher" first steamed out on the rails at Killingsworth, England, and still another decade passed before the first locomotive was placed on a public railway. It was Stephenson's "Locomotion" which opened service on the Stockton and Darlington Railway in 1825. The "Stourbridge Lion" discussed by Jervis is very similar in construction to "Locomotion," but it was built by Stephenson's competitor, John V. Rastrick.

The first American actually to witness a steam locomotive in operation was probably William Strickland, who visited England's Stockton and Darlington Railway in September, 1825, on behalf of the Pennsylvania Society for the Promotion of Internal Improvement. The question of "firsts" with respect to American locomotive development is somewhat debatable, but the following summary may help to put Jervis' contributions in perspective.

The first steam locomotive to operate in America was Rastrick's "Lion" which was driven by Horatio Allen on August 8, 1829, on the tracks of the D&H Company at Honesdale, Pennsylvania. It was never put in regular service.

The first American-built steam locomotive to operate in America was Peter Cooper's "Tom Thumb," which was given its first trial on August 28, 1830, on the tracks of the Baltimore and Ohio Railway Company at Baltimore. It also was never put in regular service.

1. See *Coalboats to Tidewater,* by Manville B. Wakefield (South Fallsburg, N.Y.: Steingart Associates, 1965).

The first American-built steam locomotive to be put in regular operation in America was the "Best Friend of Charleston" which was test operated during November and December of 1830, and hauled paying passengers on December 25, 1830. Its designers were Horatio Allen, Charles E. Detmold, and E. L. Miller, and it was built by the West Point Foundry in New York City. It exploded on June 17, 1831.

The first successful steam locomotive which operated in America was "DeWitt Clinton" of the Mohawk and Hudson Railroad.[2] It started service on August 31, 1831, at Albany, New York. It was the third locomotive built by the West Point Foundry, and it operated until 1848. However, the "West Point," the second locomotive built by the West Point Foundry, seems to have been put in service in March, 1831—almost six months before the "DeWitt Clinton." How successful the "West Point" was remains a moot question.

On the question of who invented the locomotive truck, it would seem that Jervis is now acknowledged. For example, John H. White states flatly, "It was John B. Jervis who is generally credited with first applying the truck to the locomotive. His design was developed in 1831–32. Its merits quickly became apparent, and by 1835 it had been universally recognized in this country."[3] Another recent writer, Roger Burlingame, also gives Jervis full credit.[4]

The importance of Jervis' invention may be illustrated by this excerpt from the presidential address of the American Society of Civil Engineers, May 25, 1880:

> Not only do our [U.S.] locomotives pull greater trains than do European in proportion to their own weight, but they run more miles in the course of the year . . . This has been accomplished by a series of improvements in . . . our locomotives . . . which fairly entitle us to speak of them as American in design. These improvements chiefly consist: first, in the substitution for a

2. G. H. Thompson, *Transactions*, ASCE, XXIII (July, 1890), 44.

3. *Introduction of the Locomotive Safety Truck*, Bulletin of the U.S. National Museum (Washington, D.C.: Government Printing Office, 1961).

4. *Technology of Western Civilization*, ed. by Kranzberg and Pursell (New York: Oxford University Press, 1962), p. 429.

rigidly attached pair of leading wheels, of a four-wheeled truck pivoted at its centre.[5]

The author of this address, Octave Chanute, was not only a leading civil engineer of his day, but also the brilliant pioneer in aeronautics on whose work the Wright brothers based their aircraft designs.

The Mohawk and Hudson Railroad was originally a seventeen-mile line from Albany to Schenectady. It was the first chartered railroad in the United States—on April 17, 1826—to develop into a successful operation with steam locomotives. Two other railroads, however, actually had steam locomotives in operation earlier. The Baltimore and Ohio Railroad, chartered February 28, 1828, put its first steam locomotive (the "York") in service during July, 1831, though horse-drawn service had been initiated over a year earlier on May 21, 1830. The Charleston and Hamburg Railroad, chartered in January, 1828, first utilized steam locomotion (the "Best Friend") in December, 1830. The chief engineer of this company was Jervis' friend Horatio Allen, who completed this line in October, 1833. It was one hundred and thirty-six miles long, the longest in the world when completed.

The Mohawk and Hudson was the nucleus of the New York Central System (now Penn-Central) which was formed as a consolidation of ten short lines on April 17, 1853. At that time, it was the longest single system in the world, about three hundred miles of through service from Albany to Buffalo and Niagara Falls.

Jervis mentions the lack of enthusiasm displayed by local citizens for the embryonic railroad. Local citizens were propagandized by freight-wagon, stagecoach, and canal interests to the extent that feelings ran high against the railroad. Financing was also a major problem, four years elapsing between the opening of company books and the beginning of construction of the railroad. In addition to the railway itself, the company also built docks and a warehouse on the Hudson at Albany and a special basin for Erie Canal boats at Schenectady.

The railroad had other problems. The canal route was thirty

5. *Transactions*, ASCE, IX (June, 1880), 228–29.

miles long and included twenty-seven locks between Albany and Schenectady as compared with the railroad's seventeen miles and one incline. However, loading and unloading at the basin was very expensive, and shippers were slow to use the railroad. The passenger service proved popular from the outset, and the company's stock commanded a premium price shortly after operations began.

During Jervis' tenure as chief engineer on the Chenango Canal, he was courting Cynthia Brayton. They were married on June 3, 1834. Coincidentally, both his marriage and the project ended sadly—his wife died from childbirth on May 14, 1839, a few days after their infant daughter. Jervis married Eliza Ruthven Coates on June 14, 1840, a little over a year after the death of his first wife. The canal was a financial failure, principally because it was born at a time when railroads were becoming competitive for freight, as well as passengers. Also, soon after it opened, the financial panic of 1837 throttled the entire country. With state aid, it continued to operate, but as was the case of other laterals to the Erie, it was abandoned during the post–Civil War era.

The main line of the Erie had a long period of prosperity and two major enlargements. Jervis was associated with the first period of enlargement; then around the turn of the century a massive renovation and enlargement took place. The reconstruction was completed in 1918 at a cost of over one hundred million dollars, with a minimum depth of twelve feet and width of seventy-five feet (channel bottom).

It has been one hundred and fifty years since Jervis joined the Erie Canal as an axeman. Now known as the New York State Barge Canal, it supports a fleet of about seven hundred craft which carry several million tons of products each year, especially grains and petroleum. Today, the deadly Montezuma swamp has been largely drained by the reconstructed canal system so that farmlands occupy the once desolate area.

Up until the time of the Croton project, there were only about nine public water supply systems in the United States, the remaining thirty-five being private. Most of them (Latrobe's Fairmont works in Philadelphia, a notable exception) were little more than large, centrally located cisterns. New York City's water supply system had an interesting history which is quite typical of our larger cities. In the earliest times of European settlement, there were private wells. Then around 1658 the Dutch dug a series of public wells. By the time of the Revolution, many of the shallow wells in the city were unsuitable, and water from upper Manhattan was imported by the hogshead. At that time, Christopher Colles proposed and a private company undertook the construction of a large, deep (greater than thirty feet) well. An imported Newcomen engine, which had proved successful in dewatering mines in Britain, was to be the power source. The war killed the project. About 1800, Aaron Burr's Manhattan Company began its career of supplying water to the desperate city. Unfortunately, the company's main interest was in banking so its waterworks suffered. The waterworks consisted of a well, two steam pumps, a reservoir, and about twenty miles of wooden pipe. In 1830, a huge well, one hundred and twelve feet deep, was sunk. From it, a steam pump drew water into a holding tank and thence it was delivered through about nine miles of cast iron pipe. But the supply, which was of inferior quality, never seemed to catch up with demand. In addition to problems of public health—for example, the cholera epidemic of 1832 and high typhoid rates—the growing city had an increasing hazard of conflagration.

By 1835 public pressure resulted in a special election and the approval of the Croton project. After initial surveys by Douglass and White, the plan was formulated; with Jervis' acceptance of the stewardship in 1836, the project commenced in earnest.

Of the Croton project, President Frederic P. Stearns of the American Society of Civil Engineers said in his annual address, on June 26, 1906: "New York, especially, displayed great courage in deciding to build an aqueduct 40 miles long." Stearns used the phrase "great courage" because the aqueduct was of almost unprecedented size. London's New River Aqueduct, for example, was only about twenty miles long. Only the great works of ancient Rome, the Aqua Marcia (144 B.C.) with its fifty-eight-mile length and the Aqua Claudia (forty-three miles; A.D. 52) exceeded the Croton. But the record length of the Croton system was short lived, for in 1847 French engineers completed a fifty-seven-mile aqueduct for the city of Marseilles.

The Old Croton was the first important modern aqueduct in the world. It is interesting to note that it supplied reasonably pure water to the city (about eighty million gallons per day) and there was no treatment involved. It was not until after the Civil War that filtration was practiced and not until 1908 that chlorine was put in service on a regular basis as a disinfectant.

Fortunately for Jervis and his colleagues, Charles S. Storrow published, in 1835, his famous *Treatise on Water Works for Conveying and Distributing Supplies of Water.* This brilliant work put at the disposal of the Croton engineers the most advanced knowledge of hydraulics that Storrow had gathered in Europe. Not only was the methodology of hydraulics largely untested in the 1830's, but even the methods of measurement were primitive. Clemens Herschel, who invented the "Venturi" meter for measuring flow, was born the year the Old Croton was finished, in 1842.

The advent of modern dams is said to have begun with the Zola Dam in France about 1843, the year after the Croton system was finished. This would place the Croton Dam at the end of the transition period, and its design would certainly warrant this position. Jervis seems to have been quite unfettered by convention when he approached the problem of damming the Croton River. The resulting structure was one of a kind: a completely unique combination of two ancient types, earth-fill and masonry. The result was an earth embankment about one hundred and thirty feet long and a composite spillway section about two hundred and seventy feet long. Before the devasting flood of January, 1841, the earth section was planned to be two hundred and ten feet long.

The composite section was basically as follows: (a) the natural river bottom of "Alluvium with Boulders" was smoothed and surface boulders removed; (b) huge rock-filled cribs of hemlock timbers and oak crossties were built on top of the river bed; (c) a 33-foot-high masonry dam was constructed over the timber cribs; (d) a triangular section of earth, 50 feet high (from the river bed) and with a base width of 275 feet, was placed against the upstream face of the masonry dam and timber cribbing.

The "reverse curve" or ogival section Jervis used is now a common shape for spillways, and the stilling basin is another feature of the original Croton Dam that became accepted practice.[6]

The New Croton Dam, which was completed in 1906 downstream from Jervis' structure, completely flooded out the original dam. The new dam was, at two hundred and thirty-eight feet, the highest masonry dam in the world at the time of its completion. Today, well over two hundred dams in the United States alone exceed this height; some are over three times higher.

Before 1937, when five of the main stone arches of the High Bridge (also called Aqueduct or Harlem Bridge) were replaced by a single steel arch, this beautiful masonry aqueduct consisted of eight main spans of 80 feet and seven side spans of 50 feet. Six of the side spans were on the North (Bronx) side. Its total length was fourteen hundred and sixty feet, the top of its parapets was one hundred and sixteen feet above mean high water, and the width at the parapet was twenty-one feet. The two thirty-six-inch conduits were placed beneath a roadway. The Harlem River at the bridge site was six hundred and twenty feet wide at flood tide, three hundred feet at ebb tide. The main channel was sixteen feet deep, but at flood, the tidal flats were covered to a depth of four to five feet.

The outstanding feature of High Bridge was not readily observed. Unlike most stone arches, the loads from above the arch ring (actually, an arch slab) were carried to the ring by a series of masonry walls instead of the usual earth fill. Thus, there was a great hollow space between the sidewalls (spandrels) of the arches, and the consequent "dead load" was substantially reduced, permitting a more economical design.[7] Even so, the bridge

6. A spillway carries excess flow over a dam. The spillway water empties into a stilling basin before returning to the natural stream channel. This prevents erosion of the stream bed at the toe of the dam, and thus reduces the possibility of the dam being undermined.

7. The famous Cabin John Aqueduct of the Washington, D.C., water system, which had the longest single span in the world at the time of its completion (1857–64), also used this hollow construction.

cost somewhere between seven hundred and fifty and nine hundred and fifty thousand dollars.

Although not the grandest stone arch in the world at the time of its construction, its tall piers and great length (almost six hundred feet longer than the famous Pont-du-Gard) won international acclaim for Jervis. And yet Jervis, as the engineer, had strongly advocated a far less spectacular, but much more economical low-level crossing.

The unfortunate situation which resulted from Jervis' replacement of Douglass as chief engineer of the Croton project developed into a highly personal controversy which lasted twenty-five years after the death of Douglass. Jervis presents his side of the affair in great detail. He is very careful to describe his innocence regarding an intrigue to succeed Douglass.

For comparison, a pro-Douglass presentation of the controversy is found in *Campaigns of the War of 1812–15*, written by a distinguished military engineer, Brevet Major General George W. Cullum. It was published in New York by James Miller in 1879. Some of General Cullum's innuendoes may seem biased, but his great admiration for Douglass is understandable.

> During the winter [1835], Douglass was engaged in office work which brought him in frequent contact with the Commissioners, who could not or would not appreciate the scientific character of this great work; who interfered with discipline over subordinates; were unwilling to establish an engineer department and define Douglass' prerogatives; and, in fact, treated this vast undertaking as little more than an extended job of plain masonry, which might be carried out with trifling expense.
> In the Spring of 1836, Douglass, with his small force, was again in the field. . . . while [he was] thus engaged . . . the commissioners were plotting in the city. Suddenly, without making any charges against the Chief Engineer, the Commissioners passed a resolution removing Douglass, November 4, 1836, and appointed as his successor a gentleman [i.e., Jervis] with whom it appears they had been in correspondence for some time.

Later Cullum alludes to the flood of January, 1841, and implies that the destruction of the earth dam would not have happened had Douglass' plans been followed.

It seems that two highly ethical and intelligent men were drawn by circumstances into hostile feelings. Jervis the practical, once he had accepted employment, became a staunch "company man"; the other, the fiercely individualistic Douglass, seemed to believe that he always knew what was best for his employer and eschewed compromise.

Jervis' professional reputation reached an all-time high when, on October 14, 1842, all the church bells in New York rang from dawn to dusk; the firing of one hundred cannons initiated a huge parade of over a hundred units, and the governor of New York, William Seward, officially opened the Croton System.

Boston was anxious to have Jervis study its water supply problem, for he was now the preeminent authority. Actually, a a Boston civil engineer, Loammi Baldwin II, had prepared a competent plan for Boston in 1834, but he passed away before the necessary political action could be taken.

Boston was perhaps the first town in the country with a "public" water supply, dating from 1652. About one hundred and forty-four years later, a local pond was tapped because the earlier supply from springs became insufficient. The city soon overwhelmed the resources of the pond, and for almost half a century the city could not decide on a specific course of action. The Croton may have provided the final catalytic action. With Jervis as consultant and Ellis Sylvester Chesbrough as principal engineer, the work was successfully completed in 1848. Chesbrough served Boston as city engineer until 1855 when he accepted an offer from Chicago. He was elected president of the American Society of Civil Engineers in 1878, which, coincidentally, was the year Boston completed a new aqueduct from the Sudbury River to serve the expanding population.

The beginnings of the New York Central Railroad were mentioned earlier. The Hudson River Railroad which Jervis engineered remained an independent short line until 1862, when the crafty financier, Cornelius Vanderbilt, acquired it through stock manipulations. But this was only one step in the Commodore's master plan.

Taking advantage of the fact that the New York Central used Hudson River steamboats for the bulk of its transshipments during the ice-free season and only then used the Hudson River Railroad, Vanderbilt forced the price of the Central's stock down and on December 11, 1867, bought controlling interest.

Jervis' last major services as an engineer were all associated with railroad connections to Chicago. He asserted that financing was the main difficulty encountered in the Michigan Southern and Northern Indiana Railroad. This may have been true in 1850. However, between then and 1855, six major linkages with Chicago were completed to the east-coast cities of Boston, New York, Philadelphia, Baltimore, Norfolk, and Charleston.

Although Jervis was president of the Chicago and Rock Island Railway during its construction, Stewart H. Holbrook, in *The Story of American Railroads,*[8] gives full credit for the project to Henry Farnum, Jervis' younger colleague. Farnum had first worked with Jervis on the Erie Canal and later on the Michigan Southern Railway. He succeeded Jervis as president and was responsible for the first railroad bridge across the Mississippi River. It was this bridge that created a major legal battle (the "Effie Afton" case) that involved a promising young attorney named Abe Lincoln. The litigation concerned allegations by steamboat interests that the bridge's piers were a hazard. Lincoln's adroit handling of the case for the bridge company enhanced his national reputation as a lawyer even though the issue was not decided until an 1862 ruling by the United States Supreme Court, five years after Lincoln left the case with a hung jury. It was Jervis who, acting as a consultant on the bridge, wrote the basic report (see Chronology of Jervis' Professional Life for 1857 pp. 189–90).

Jervis' last project was undertaken when he was sixty-six years old, and unlike all his earlier work, it was principally operating rather than building a system. This was the Pittsburgh, Fort Wayne and Chicago Railway. In 1856, the renowned railroad executive and one-time engineer, J. Edgar Thomson, had promoted the consolidation of many short lines into this system and in 1869, he, as president of the Pennsylvania Railroad, leased and

8. (New York: Crown Publishers, 1947).

finally bought it. In his recounting of the success of the Pittsburgh, Fort Wayne and Chicago Railway under his stewardship, Jervis perhaps should have mentioned the fact that the Civil War was at least somewhat responsible for the high revenues.

Jervis' uncomplicated view of right and wrong and his simple philosophy of success is clearly manifested in his discussion of his profession. In *Labor and Capital,* a book which may have been written about the same time he wrote the chapter "Professionalism and the Engineer," he goes to great lengths to extol the pioneer virtues of hard work and honest living. He was his own best example—a poor farm boy who became the "Nestor of American Engineers."

Two major activities seemed to have dominated John Jervis' life: his profession and his religion. His minor excursions into politics—such as running for state engineer on the Democratic ticket in 1855—were more or less extensions of his professional philosophy of how to get things done. Only passing mention is made of his religious views in the body of this work, but because they were such a great part of his life, they will be discussed here.

Jervis was a man of independent nature, and though immersed in frontier Calvinism from birth, he sought his own rationale before accepting the faith of his fathers. During his early manhood, he seems to have been particularly concerned about apparent contradictions between science and religion. However, he was ultimately to live by those firm religious convictions which were derived from his Protestant upbringing through an almost continuous process of evaluation and contemplation.

The quiet confidence that Jervis found in his personal religious views did not seem to engender a supercilious or missionary complex in his character. All evidence indicates he was a very tolerant man. This may have been, in part, built-in by the nature of his own church in Rome, New York, the First Religious Society, which specifically welcomed all faiths ("and Indians") to worship at their services.

Although founded in 1793, the Society did not have a permanent pastor (the Reverend Moses Gillet) until 1807 or a permanent church building until 1808. Jervis and this church grew up together. He formally joined the church when he was twenty and was among those who established a separate Congregational Church in 1834 but were later reunited with the First Religious Society. Still later, however, the Society accepted the Presbyterian Polity, and Jervis was elected a member of the Board of Sessions in 1866. He served for many years as an officer of the church and faithfully attended services to the end of his life, even though he had to come by wheelchair in his last years.

Some indication of Jervis' intellectual fascination with religion is found in his library which had over a thousand books. In it were over two hundred and sixty works on religion and related subjects. Typical of the many titles are *A Dissertation on the Prophecies* (G. S. Faber, 1811), *A Comparative View of Christianity and of Other Forms of Religion* (W. L. Brown, 1829), *The Way to Do Good* (Jacob Abbott, 1836), and *The History of the English Bible* (B. F. Westcott, 1868). In addition to books, many religious periodicals were in his library.

His contributions to religious activities were many and varied. He seems to have been first in line when his church in Rome needed funds, but he also helped along other churches and, especially, missionary groups. For example, in 1834, he gave one hundred dollars to support the Presbyterian Church in Honesdale, Pennsylvania. In his will are also found a series of religious bequests. After providing for his wife, Eliza, and his sister, Elizabeth, he divided the remander of his estate's funds among four Protestant groups and two schools: American Board of Foreign Missions, American Bible Society, Presbyterian Home Missionary Board, American Missionary Association, nearby Hamilton College, and the Normal Agricultural Institute of Hampton, Vermont.

Perhaps the most revealing part of his will is found in the section relating to the use of the library he was establishing for the "liberal-minded citizens of Rome." Here he says:

> The audience room shall be used for any secular or religious use on any day of the week from Monday to Saturday that is not

of immoral tendency. On the Sabbath Day or Sunday it shall only be used for the worship of God and the teaching of Christian doctrine according to Christian scriptures. In this, preference shall be given to such individual or Society as hold and teach the following doctrines with such as are not inconsistent with the same, viz: that God is and is the creator and preserver and governor of all things; second, that Jesus Christ came into the world to save sinners; third, that God, by his mercy and grace, saves sinners through the atoning sacrifice of the Lord Jesus Christ; fourth, that God rewards and sanctifies all sinners that are saved; fifth, that the ethical principles shall be founded on the Decalogue or ten commandments as illustrated, enlarged and spiritualized by the Lord Jesus Christ in his Sermon on the Mount as recorded by the Apostle Matthew; sixth, that God has appointed a day in which he will judge the world in righteousness according to the deeds done in the body. The result of which will be that the wicked shall go into everlasting punishment and the righteous into life eternal.

When applying these tenets to real-life situations, however, Jervis seems to have adapted them into a more philosophical form such as when he wrote a bereaved friend: "I shall not attempt to offer anything by way of consolation; too well aware it is one of those events that time alone . . . can console, but, my dear friend, this is the way we all must go. We witness the departure of near and affectionate friends, and soon in time shall ourselves be the subjects for others to mourn."

John Jervis passed away on January 12, 1885, in the city of Rome, New York, where he had arrived as a small child four score and seven years before. His grave is marked with a simple gray granite shaft atop a shady knoll.

His life was an outstanding example of the American Dream. He rose from a lowly axeman on a survey party to become an international figure in the profession of civil engineering—the leader of his profession in America. He had joined it in its infancy and helped it grow to lusty manhood. He and his colleagues were the link between the founders of civil engineering in America—Benjamin Wright, Benjamin Latrobe, and Loammi Baldwin—and the sophisticated generation of university-trained engineers who carried its traditions up to the threshold of the twentieth century. The traditions were of integrity, hard work, and achievement—traditions wrought by men like John Bloomfield Jervis.

THE REMINISCENCES OF
JOHN B. JERVIS

ENGINEER OF THE OLD CROTON

I

EARLY LIFE

ACCORDING TO the family record, I was born at Huntington, Long Island, New York, on the fourteenth of December, 1795.

My father was Timothy Jervis of the same place, and my mother Phebe Bloomfield of Woodbridge, New Jersey.

My father was a carpenter by trade and in the latter part of his life was occupied mostly in lumbering and farming.

In 1798 my father moved to the town of Rome, then known as Fort Stanwix. This place was chiefly prominent as having a navigable canal[1] of about two miles in length, connecting the Mohawk River with Wood Creek, and by connecting improvements by means of locks and dams on the Mohawk River and Wood Creek, which were parts of a system of improvements to connect natural navigation, through a large portion of the State of New York. By these means, barges or batteau boats made navigation from Schenectady on the Mohawk to Ithaca at the head of Cayuga Lake. The boats were often of ten to fifteen tons burden, worked by setting poles, oars, and sails.

The country at this time was mostly a wilderness of heavy timber. The site of the present city of Rome was then for the most part covered with native woods.

I was the eldest of seven children,[2] and had the experience of the trials of settlement in a new country.

1. This navigable canal was the work of William Weston, a famous English engineer who aided the Western Lock and Navigation Company in this project. It was one of the earliest American lock canals, built in 1795; others were built around the Great Falls of the Potomac River near Washington, D.C., and on the Connecticut River near South Hadley and Turner's Falls.

2. The others were Ann Eliza, Susannah Maria, Elizabeth, Timothy Brewster, William, and Benjamin Franklin.

I was kept at the common schools of the day most of the time until I was fifteen years of age. There was then no public school sustained in whole or in part by the state. After the close of my school time, I was occupied with my father until I was twenty-two years of age. During these seven years I was unsettled as to any definite vocation. Various plans were discussed from time to time, but nothing settled. I was handy in managing a team, and though of light frame, I was soon able to tend a sawmill which constituted my principal work during the summer, with occasional farm work; that of hauling saw logs and wood was my occupation in the winter.

My parents were members of the Congregational Church, the first religious order organized in the town. The first installed pastor was the Reverend Moses Gillet, from Connecticut. I connected myself with this church when I was about sixteen years of age. At that day, one of the important aids in religious education was the old New England primer. My parents were extremely anxious for what they regarded likely to promote the happiness of their children, and next to the Bible, the primer was the guide to religious education. I may remark, this primer after a time was neglected, and I had not seen a copy for many years when I saw a notice of a new edition published at Monongahela City, Pennsylvania. I sent for several copies, and found it essentially a reproduction of the old primer. It was Calvinistic in theology, though it did not enter into the Calvinistic policy of church government. It was an excellent epitome of scripture teaching, laying the foundation of evangelical doctrine.

Pastor Gillet was a Calvinist, though not holding the erroneous views which have been attributed to Calvin by his enemies. He was a plain, sensible man, and his teaching was essentially Biblical. Under it, his church grew in the knowledge of the scripture, and became a body of intelligent Christians. Under this kind of teaching, I adopted as my religious sentiments the Calvinistic theology, except as to church polity. This was my education, and as I advanced in life, my experience and observation confirmed my early sentiments, and now I regard *Calvin Insti-*

tutes [3] as a sober and eminent production. In this I do not assent to every particular, but as a whole, regard them as eminently Biblical.

I mention the above not to claim that I have exemplified those sentiments in all respects, but to show the basis on which I started in life; a basis that has sustained me through many trials and embarrassments, and by which my affairs have been so conducted that I am not conscious of having injured any man or, in any case, exercised unnecessary severity. My reader may say, "He was rather puritanical and perhaps trusted in God, when he should have been at work." To this I reply that a proper trust in God does not exclude the means God has provided for our use. It rather inculcates prudence and energy in conforming to the conditions imposed in the rule of obedience to His commands. I have been somewhat particular in setting forth my early training, as I consider it the basis of my subsequent life. All through my days it has been a monitor and reproof of my wrongdoing and a supporting pillar to whatever I have done that is good.

In subsequent life, my path has often been crossed by skeptical views in regard to religion. Natural law has been claimed as the potency of life, and various views have been presented to sustain the theory. I could not change the reign of law, nor could I see how law could operate on what did not exist, and hence there appeared a necessity for some antecedent to law—some power to bring a thing into subjection to the law, under a purpose that should provide for its ultimate development and use. So I fell back to the idea of a personal God.

At one time I entertained the idea of obtaining an education of a higher order than I could have in common school. But my father was not able to help me, and in fact, he needed the service I could render him. I thought to study and work at the same time, and made arrangement with an educated teacher to hear my recitations, and began the study of Latin. The time I had to study was—after setting the log, I could study until admonished

3. *Institutes of the Christian Religion* was written by John Calvin in 1536 as a guide on doctrinal matters for his Protestant followers.

of the time to reset the log. After making what I considered a fair experiment in this method of obtaining an education, I gave it up. The demands of the work I had to do absorbed so much of my time and thought that I felt compelled to give up the attempt.

I may here remark that I regard it a great error to bring up a boy without giving him the training of a regular vocation. But my home was pleasant, and my father needed my services; and though he manifested anxiety in regard to my future, he could not overcome the urgency of my usefulness with him.

As an instance of the times, I mention that in the winter season I engaged to transport a load of wheat for a merchant in town from Rome to Albany [about one hundred and ten miles]. It was good sleighing until I passed Schenectady when I found the snow rather light. I took with me my provisions and oats for my horses. It required seven-days' journey. I received for the service about eleven dollars. My expenses were about four dollars, besides provisions and oats for my horses. The latter at that day were cheap, worth together about three dollars, leaving me for myself and team, about four dollars for seven-days' work. This is the way we transported in those days, a circumstance that impressed upon me the importance of the canal, which soon after was presented to the public mind.

Though but a youth at that time, I regarded the canal project with much interest. I had no definite idea as to whether it would pay for construction, but somehow I thought it a great work. I had not the least idea of anything personal, more than to supply such demands as it might make for lumber. The subject was very warmly discussed, and many intelligent men regarded it as a measure that would involve the state in inevitable pecuniary ruin.

In 1816 the state made an appropriation for an instrumental survey and estimate of the Erie and also the Champlain Canal. With some previous surveys the estimates were made by the Honorable Benjamin Wright [4] of Rome, the Honorable James

4. Benjamin Wright was Jervis' mentor and life-long friend. Born in Wethersfield, Connecticut, of old New England stock, Wright rose from a local surveyor in Fort Stanwix to chief engineer of the Erie Canal. Later he directed major

Geddes [5] of Onondaga, and John Broadhead [6] Esq. of Utica for the Erie Canal, and by Mr. G. Lewis Garin for the Champlain Canal. With the information furnished by these engineers, the state entered on the construction of the works. Mr. Wright was appointed chief engineer of the Erie and Mr. Geddes of the Champlain Canal.

Several sections of the canal between Rome and Utica were put under contract, and the formal ceremony of breaking ground occurred on July 4, 1817. There had been a difference of opinion on the question of location through Rome, in consequence of which it was delayed until late in the autumn. The general principle of location having been settled, a party of engineers was sent on to make the survey through the Rome swamp.

Mr. Wright, the chief engineer, as above noticed, resided in Rome. When the party came to make the survey, it was composed of only the experts, depending upon obtaining axemen at the locality. These latter were an important adjunct in carrying a line through a dense cedar swamp. To supply this defect in the engineering party, Mr. Wright, the chief engineer, knowing that my father had in his employ men working as axemen, called upon him to engage two for a few days on this service. My father referred the matter to me; he was induced to do this from an intention of making proposals for a contract for this section of the canal, and thought I would be able to obtain useful information as to the character of the work. I assented to his proposition to go as one of the axemen myself, taking with me one of my father's men.

Having been brought up in a new country, I had acquired a

---

canals in New York, Maryland, and Virginia, railroads in New York and Cuba, and many other public works. He has been called the Father of American Civil Engineering.

5. James Geddes came to New York from Pennsylvania. He studied law and was a judge and surveyor who, as early as 1808, made surveys for the Erie Canal. Later he was chief engineer on the Champlain Canal. He also worked on canals in Ohio, Maine, and Maryland.

6. Charles C. Brodhead, also Broadhead, came to Utica from Kingston, where he had been a local surveyor. He had surveyed the Black River area in 1793. Why Jervis calls him John is uncertain.

very handy use of the axe, and my associate was a first-rate chopper. As a brace of axemen, we entered on this new field with good spirits, not doubting we could do anything required in this line; and beyond this I had no thoughts except to notice the prospect of a contract for this swamp section of about two miles. We had had no experience in cutting lines for such a purpose. The cedar brush was very thick, and the ground very soft, and to keep on a line was a thing we had not practiced. But we soon got the run of the work, and having myself a pretty good eye for a line, we worked our way in what appeared a satisfactory method. The swamp was so soft, it was necessary to drive stakes to set the leveling instrument on. In our progress (which was in November) a fall of snow occurred, loading the cedar brush so as to render the work for the axemen very disagreeable. It was necessary that they should go forward, and by cutting shake the snow from the brush and trees, and so open a path for the engineers.

I was of a slender frame, rarely weighing over one hundred and twenty-five pounds; my associate was tall and stout, and we pushed through the brush and snow with the zeal of a new enterprise. Our main ambition was to satisfy the chief of the party, N. S. Roberts Esq.,[7] who was a man of austere manners who did not hesitate to speak plainly.

Though my occupation in this service was that of an axeman in which I exerted myself to cut stakes, pegs, and line in a satisfactory manner, I had not proceeded long before my attention was attracted to the operations of the men using the instruments. These instruments and the method of using them appeared to me a profound mystery, and I at first had no idea, considering my education, that I could learn their mysteries. The two target men, for whom I drove the pegs on which they rested their instrument for an observation from the level, came often to my position, and after driving their pegs, I had some leisure until the

7. Nathan S. Roberts was of Pilgrim stock. A part-time school teacher from Plainfield, New Jersey, he arrived at Oriskany in 1806 to practice this profession. It was Wright who persuaded Roberts to be a surveyor and engineer on the Erie. Later Roberts worked on the Chesapeake and Ohio Canal and was chief engineer of the Pennsylvania State Canal.

observation was made to notice their manipulations. On such occasions I was led, at first from mere curiosity, to watch their movements and to ask some questions, which I did with great modesty, as they quite outranked me in the service. After a time, as I became more familiar, and, having some approving words for my expertness as an axeman, I ventured at some leisure time to ask the privilege of taking the target and of examining its movements, which was kindly conceded. After a day or two, I ventured to ask some questions in relation to this mysterious instrument and the mode of its operation. I learned in these conversations that the target man was regarded as occupying the first step in the science of engineering.[8] On reflection it appeared to me I could do that service if I could have a little practice.

The chief of the party, Mr. Roberts, was, as I have observed, a man of austere manners, and the party generally regarded him with a reverence that did not allow familiarity. In the low rank of my place in the party, I did not feel at liberty to ask of my principal any questions, and all my communications were with the target men, who were both young. For a time I did not see my way clear to bring out the working of my own mind; but by and by the principal seemed to unbend a bit, and on an occasion expressed his approbation of our dexterity as axemen, and very frequently stated that he had never had a pair of axemen so skillful and efficient as we were; and at the same time took occasion to manifest a partiality to men who were not afraid to work. This encouraged me to think that the same application of industry might enable me to master an art that seemed then so mysterious. Gathering strength by reflection, the idea I had long indulged so pressed on my mind that on the last day of this service, while the party gathered in a huddle in the swamp and were eating our dinners, I mustered courage to make a proposition, partly in joke and partly in earnest. I put it in the modest form of a question to Mr. Roberts: "What will you give me to go with you

---

8. The survey party was and still is generally organized as follows: party chief, instrument man, chainmen (front and rear), targetmen (or rodmen), axemen, and helpers. A potential surveyor would advance through each of these steps to party chief and thence perhaps to resident engineer.

next summer and carry one of those targets?" He very promptly replied, "Twelve dollars per month." [9] I as promptly assented and said I would go.

As this seemed to settle the question, I began to consider seriously what it was that I had given my assent to. It was no object to engage my services for twelve dollars per month. I must see something beyond this. It was natural the question should arise, could I ever become an engineer? The thing appeared a mystery to me, of which I had only seen a very small edge of the great field. In this course of reflection I considered first that I only had a common school education with no view to the science of engineering. So far I had only looked at the target; the mystery of the level, the taking of sights, its adjustment, and the computations of the observations were all dark to me. As I pondered these matters, I began to think that after all, it might be a mere joke and that Mr. Roberts had no idea of my accepting his prompt assent. But the idea held fast to my mind, and finally I determined to ascertain if the parties were in earnest. I say parties, for Judge Wright [10] was the principal, and I had intimated to him Mr. Roberts' acceptance, to which he did not object. Having a personal acquaintance with Judge Wright, the chief engineer, after reflecting several weeks on the subject, I called on him and requested him to inform me if they certainly accepted my proposition. He replied very promptly, "If you say you will go, I say you shall go."

Now a new occupation suddenly burst on my view. How shall I prepare for it? Only a few months and I shall be called upon to enter a party, all probably better educated than myself. I had no fancy for appearing more ignorant than those engaged in the same field. The whole aspect appeared to me a profound mystery. But I had voluntarily put on the harness, and had faith to believe that what others had learned, I could learn.

On inquiry I found that land surveyors were considered best

9. The engineers, by comparison, received from about one hundred to one-hundred and fifty dollars per month.

10. Wright was active in local politics. He served as a state legislator and, during the War of 1812, was appointed an Oneida County judge.

fitted for engineers. So, at the recommendation of Mr. Roberts, I purchased two books on surveying.[11] These I studied in the evenings, and at such odd times as the weather did not allow team work, until 10th of April 1818, when I started with my target on my shoulder, with a party of about a dozen men for Geddesburg (now Syracuse). Except N. S. Roberts, engineer, the principal of the party, we all made the journey on foot. A baggage wagon accompanied us, with our baggage and camp furniture. We started in the afternoon and only made seven miles that day. The roads were excessively muddy, and we reached Geddesburg in the afternoon of the third day and then pitched our tents among hemlock logs and bushes. On the journey, I occupied such opportunity as I found in studying the target, making myself familiar with the movements, as I had seen in the swamp, and especially with the reading of the figures. The target rod used at that time was ten and a half feet long.[12]

My neighbors at home laughed at my venture, and predicted I would soon return home to sleep.

On arriving at Geddesburg, the party immediately organized for work, Mr. Roberts having arrived about the same time and joined the camp. The party then proceeded on the work of locating the canal from this point to Montezuma [13] on the Seneca River, a distance of about thirty-six miles.

A few days after we began work, a severe snow storm came on, falling to a depth of six inches, and followed by high winds not agreeable for tent life. As the season advanced, the warm

11. J. Day's *Principles of Navigation and Surveying* (New Haven, Connecticut) became available in 1817. Other common books on surveying were Abel Flint's *System of Geometry and Trigonometry, Together with a Treatise on Surveying* (Hartford, Connecticut, 1804), and S. Morris' *An Accurate System of Surveying* (Litchfield, Connecticut, 1796). It is not known, however, which books Jervis bought at this time.

12. Modern rods are about six and a half feet long, and can be extended to twelve to thirteen feet.

13. In the 1790's possibly the longest wooden trestle bridge in the world (over two hundred spans at twenty-five feet) was built over the Seneca River near here. The swampy areas which necessitated the long bridge also caused many problems for the canal builders. It is surprising that Jervis did not mention this remarkable bridge.

weather brought out large quantities of mosquitos,[14] which proved the most serious annoyance to our comfort, especially in their night visits.

In our operations it was common to walk from one to five miles to and from our camp to work. Our principal walked with us, and the task was regarded a necessary and proper thing. At the present day this would be thought rather severe service, especially as the day's work on the line was a continuous labor with half an hour for dinner.

At times, when in the opinion of the chief of the party it was necessary, our young aspirants for the profession of engineering were required to take an axe and cut pegs and stakes and clear line. Mostly they did not relish this, regarding it an infringement on their dignity, doing their work in a hesitating and slovenly manner. As I had been brought up to work, and did not regard any honest labor a degradation, I performed my part on such occasions with the same alacrity as my regular work. My readiness on all such occasions gave me favor with my principal, which I improved to obtain knowledge in the art I sought.

The target men were allowed and requested to keep a field book of the observations and work out the levels, so we had opportunity to become familiar with the method of taking the observations and the principle of computing the levels. No one was deemed competent to work the levels except Principal Roberts, and we could do this only for our own practice, but after a time, when Mr. Roberts was occupied in camp on his plans and calculations, he would allow us to practice with the level. I eagerly embraced such opportunities.

I had prepared myself with a very plain set of drawing instruments and on rainy days employed myself in plotting the line and the profile, making such a map as my limited skill allowed me. These were of no business value except for my own improvement. Mr. Roberts made all the maps that were regarded as necessary for the business. This made me familiar with the art

14. Later, many construction workers were to die from fevers carried by mosquitoes in this area. In 1820, for example, about a thousand workers were disabled from sickness between July and October.

of plotting lines and profiles, which then appeared to me the great things to be learned.

There were sometimes errors by the target men in calling off the observation. This would be discovered on running the proof line, which was always carefully done. But the discovery often led to the necessity of making the second proof. An error of this kind did not fail to call out severe reproof from the Principal. As there were two target men, it was ordinarily impossible to decide which committed the error, and so both were compelled to feel the reproof. Sometimes I thought my associate was rather punished over my shoulder; but it could not be certain on which the fault properly rested and we were both cautioned against similar error. Sometimes it was finally ascertained the fault was not with the target men at all. All this was good discipline to impress on our minds the strict care required in conducting levels —much more important on canals than on rail or other roads. Principal Roberts had a profound sense of his responsibility in establishing a correct level for the contemplated canal, and this very naturally impressed itself on my mind.

The location for the canal was completed to the east shore of the Seneca River, at Montezuma, about the tenth of July, three months from the day we left Rome, traveling on foot about eighty miles.

At Rome, the party was broken up and reorganized into several small parties that were assigned to the charge of staking out and superintending the works. At this time, or very soon after, the whole of the middle section was placed under contract.

In this new organization, one target man was considered sufficient for a party. The party to which I was assigned was placed under David S. Bates as principal, usually known as Judge Bates.[15] I had seen Judge Bates, but had no special acquaintance

15. David Stanhope Bates, son of a Revolutionary War officer, was trained as a minister but took up surveying because he loved mathematics. In 1810, he left his native Parsippan, New Jersey, for Oneida County, New York, where he was first a surveyor, then superintendent of the local iron works, and later, county judge. He joined Wright on the Erie in 1817 as an assistant engineer and advanced to division engineer. Later he was the principal engineer of the canal system for the State of Ohio.

with him. His chief qualification was as a land surveyor of good standing. I was the only one of the old party that was attached to Judge Bates, to whom was assigned the position of resident engineer of a division extending from Canastota in Madison County to Limestone Creek in Onondaga County, about seventeen miles in length.

This opened quite a new field of experience to me. I soon found that Judge Bates had very little experience in the use of leveling instruments, or the computation involved in conducting levels, and was very ready to avail himself of what I had learned.[16] He was a man of very pleasant manner, and rather to my surprise, was ready to learn, even from me. This was extremely favorable to my progress, as he allowed me to use the instruments freely, and very soon came to depend on me for taking the levels and as a check to the computations. I very readily availed myself of this privilege, by which I became very familiar with the use of all the instruments, and the calculations required in laying out and computing the contents of the work. The latter, as the surface of the country was pretty level, did not require high mathematical science, though more perhaps in some cases than was applied.

The work we had to do consisted mainly in laying out curves on the angles of the line and setting the side stakes to guide the workmen. We went along and for several days had no side stakes to set, except when the surface of the ground was above the bottom level of the canal. This was simple, as we only had to add the slope to the half-width of the bottom of the canal, the measure being from center line. After a few days we came to a piece of work where the surface of the ground was about two feet below bottom level. Mr. Bates was at the level, and I was carrying the target. The observation and calculation showed the ground below level. Considering that the same slope would be carried in from bottom level, I deducted the slope from the dis-

16. Most early surveyors were trained for compass and chain type surveys and not in leveling. That is, they knew how to make a "plan" of a piece of property but could not make a topographical map of the land as this required measuring differences in elevations ("levels").

tance to center, and so called out my calculation to Judge Bates as showing where the stake should be set. On this, Judge Bates directed the stake set at fourteen feet from center, not deducting the slope. As it was a new case, I carefully looked over the matter on which, it appeared to me, I was right. Judge Bates could not see the propriety of putting any stake less than fourteen feet from center, and so we were at issue. We pondered over the question for a while, and then left for dinner. At dinner, we were joined by Canvass White,[17] the principal assistant engineer, who was reviewing the line generally. We submitted the question to him, and after dinner he went out with us, and taking a seat on a log, appeared to us as making an examination of the question. Whether he was or not, we did not know, for after sitting about an hour, he left without giving us his opinion. We thought he had concluded it was better to let us work it out for ourselves, thinking that one of us was right and that we should ultimately reach a correct decision. I then drew sketches to illustrate my position, and toward evening we came to what we agreed was the true solution of the question. I mentioned these circumstances as historic of the science as held at that day, though it must not be understood there were none to comprehend such questions.

By the preceding, it will be noticed my situation was favorably advanced, especially as to acquiring knowledge. I received no increase of compensation, for I was technically the target man at twelve dollars per month. I was, however, well satisfied to bear all the responsibility for the privilege I had of gaining a better knowledge for the profession, which I now looked upon with great interest. In this new position, I derived great benefit from the discipline I had had under Mr. Roberts, by which I was impressed with the necessity of great care in observations and computations. Our party concluded its season's work in De-

17. Canvass White was, like Jervis, a local farm boy who became an engineer under Wright's mentorship. He joined the canal staff in 1816 and was selected to tour English canals in 1817. Returning to New York, he became a close engineering aide to Wright and achieved much fame for discovering a natural cement in Chittenango, New York, near the route of the Erie. Later White worked on canals in Connecticut, Pennsylvania, and New Jersey.

cember and returned to Rome, the headquarters of the chief engineer. Here we were disbanded for the season. I may remark, very little work was done at that day on the canal in winter. I had no idea of any employment for the remainder of the winter, but I soon found I was assigned to the duty of weighing stone for locks.[18] For this purpose, I took up my winter quarters about midway between Onondaga Hollow and Syracuse, the latter then known as Crossets Corners. This service I completed about the first of April. As a feature of the times, I will state how young aspirants for engineering traveled in those days. I accidentally met at Crossets Corners two of my associates of Judge Roberts' party of the previous summer, who were residents of Rome and who had spent the winter in the same work of weighing stone further west. There were Messrs. Barrett and Tibbits.[19] We were forty miles from Rome; it was late in the afternoon, and we decided to go on to Orville, a village four miles east, and lodge there, with the view of making Rome, thirty-six miles from Orville, the evening of the next day. Accordingly, we started from Orville the next morning at sunrise, each carrying the bundle of clothes that had served our changes for the winter. The morning was bright and the ground hard frozen, and we walked fifteen miles to Canaseraga and stopped for breakfast. After breakfast, at about ten o'clock, we resumed our journey. The clear sun had now so far dissolved the frost as to make about an inch of soft mud, materially injuring the walking. Shortly after we had resumed our journey, a stagecoach came along. There was plenty of room in the coach for us all, but not the least intimation of a desire to avail oneself of its benefits was made by any of us, and we allowed it to pass on as a thing in no way suitable for us, at least not in accordance with our independent method of traveling. When we arrived at a point twelve miles from Rome (Oneida Castle) we found the ice ridge made by the sleigh track was firm, and the valleys between nearly filled with water

18. Masonry contractors were paid on the basis of the amount of stone actually placed. The stone-weigher determined the basis for these payments.

19. Presumably Jervis is referring to Alfred Barrett, who was later chief engineer on the enlargement of the Erie, and Hiram Tibbits, a surveyor who served the Erie at least until 1823.

from the melting snow. The condition of the road, as we found
it after breakfast, made our traveling heavy and disagreeable.
We reached Rome about nine o'clock in the evening, pretty well
fatigued, but hardly felt the worse for it the next day. Not much
of this sort of service is undertaken at the present day by young
engineers.

II

RESIDENT ENGINEER ON
THE ERIE CANAL

MY FIRST YEAR closed at the conclusion of my journey home as before stated. I had not the least intimation as to the occupation I might have for the second year. Of course I was anxious on this point, and hoped at least for some position that would afford the means of improving my knowledge in engineering. I was not left many days in uncertainty before I was informed by Judge Wright, the chief engineer, that I would be placed in charge as resident engineer [1] of the division occupied the previous year by Judge Bates, he being assigned to more general duties. My position as target man for Mr. Bates made me familiar with this division of the canal. This announcement of the chief engineer gave me great pleasure, and I entered on the duties with the zeal of youth. The ordinary duties had become very familiar in the experience of the previous year. My association with Judge Bates had been of the most cordial and intimate character, and I supposed my appointment had resulted from his recommendation.

In the prosecution of this new responsibility I had no difficulty in so conducting operations as to obtain the approval of my superiors. Nothing was said to me as to the compensation I was to have, nor did I ask any questions on this point, being satisfied it was an expression in favor of my progress, which stimulated my ambition for advance in the profession.

My division had few mechanical structures. A few wood

1. The approximate ranks for professionals on the Erie were: (1) chief engineer, (2) principal engineer, (3) principal assistant engineer, and (4) assistant engineer. Resident engineer was more of a position than a rank.

trunk aqueducts, waste weirs, and bridges, and small stone culverts.[2] The plans for these were prepared in the office of the chief engineer (Judge Wright) at Rome. In what pertained to the running of lines and levels was well understood at the time, but the mechanical department of engineering was practically in its infancy. Such matters were freely discussed with intelligent mechanics whose skill was supposed to be useful in this department of engineering. The plan for a timber trunk for the aqueducts was prepared and submitted by a carpenter, Mr. Cady of Chittenango. This plan was adopted in nearly all the wood trunk aqueducts on the canal. At this day it stands as a well designed plan for this kind of structure.

Canvass White was the principal assistant engineer, and on him devolved the duty of preparing plans for the mechanical structures. Mr. White rendered a most important service to hydraulic engineering in his discovery of a material for hydraulic cement, a very important matter for canal works.[3] I well recollect his diligent examination of the stone quarries and his experiments during his search for suitable material. As usual in such cases, there was much skepticism on the part of the public until the thing was demonstrated. He was sent to England to examine

2. *Mechanical* in this sense pertains to the mechanics of physics, that is, structures designed by the laws of mechanics, for example, statics and hydraulics. This is unlike today's common usage in which *mechanical* pertains to mechanisms such as engines and the like. "Wood trunk aqueducts" were large, box-like structures resting on stone piers which carried the canal over large streams. They were usually a series of timber frames lined with heavy planking. "Waste weirs" were small spillways built in the sides of the canals to carry off excess water from canal sections to maintain the design depth of flow. Bridges were required to permit local foot or wagon traffic to cross the canal. They had either to be high enough for the canal boats to pass or to be raisable. Most of the early bridges were fixed wooden structures. "Culverts" were generally small stone arches under a canal to carry unneeded water past the canal so that it would not pond on one side of the canal, thus weakening its sides.

3. Hydraulic cement is a binding material which can set under water. At that time, most cements such as lime and plasters were leached out of masonry joints when subjected to water over a long period of time. Thus, mortars (mixtures of cement and sand) had to be constantly replaced in masonry subjected to water. Use of hydraulic cement in the mortar greatly reduced this problem. White was awarded ten thousand dollars for this find.

the canals of that country, and was the most able engineer the state had at that time, in the mechanical branch, in the engineering of her canals. Of course, all plans were submitted to the scrutiny of the chief engineer. Though Judge Wright did not claim special originality in mechanical engineering, he had an excellent judgment on plans presented, and his criticisms were very sharp in discovering their merits or defects, and all were required to pass his review.

There was no lock on my division, it being a portion of the long Rome summit of sixty miles; consequently, I had not much opportunity to acquire experience in the mechanical branch of engineering; but it was sufficient to bring to my observation the requisites of a canal in regard to what was necessary for strength of works and to provide impervious connections between the earth works and the mechanical structures.

The middle section of the Erie Canal was essentially completed at the close of 1819, of course, my division with the rest. Water was let in late in the fall on the eastern portion, and a boat ran from Utica to Rome. This was a demonstration of great joy to the people, and was made an exciting festival. I was on the boat named "Chief Engineer," [4] and I well recollect the thrill of pleasure I felt as the boat started from the wharf. It appeared to me a splendid consummation so far, and an earnest [pledge] of the vast benefit that would be realized on its completion.

Late in the fall, Judge Bates was sent with a party to make a survey of the Oneida and Oswego Rivers. I was made the surveyor and took the traverse of the lines. Snow fell before we completed the survey. When the survey was completed the party broke up. Judge Bates's residence was in the town of Mexico, Oswego County. He desired me to go home with him, and there make out the plotting of the lines. A son of Judge Bates was in the party, and we then traveled through the wild roads that led to Mexico.

When my work was completed, I was at liberty to go to my home at Rome. My route was through a new country, very sparsely settled, and I had the prospect of a lonely passage.

4. This boat was so named in honor of Wright.

Judge Bates and his son remained for a time at home. There was no public conveyance, and the journey had to be made on foot. I decided to take the route by Constantia, a small village on the north side of Oneida Lake, fifteen miles from Mexico, where my Uncle Bloomfield at that time resided. For about half the way from Mexico there was settlement, so that no long reach of woods occurred, and then an unbroken forest of seven miles, reaching to within a mile of Constantia. There was light snow on the ground. The road was cut out but little beyond the wagon track, and the sides had a thin growth of bushes over the trees that had been removed from the road. In this narrow road I did not meet a person in the whole seven miles. At Mexico, I was told of hunters who had made $500 during the year in bounties on the wolves and panthers they had killed. I certainly had a dread of the walk through the seven-mile woods, but I saw no other alternative, as the other route was nearly as wild.

I started on this walk immediately after breakfast, taking it moderately until I entered the seven-mile woods. I had no weapon but a walking stick. My eyes were on the lookout. The stillness of the forest was impressive, only disturbed by occasional falling of snow from the branches of the trees. My pace quickened as I entered the seven-mile stretch, and if my reader need assurance, I can say that the sight of Constantia was a great relief, and my thanksgiving went up to my Heavenly Father who had guarded my way through the wilderness. The next day, in the afternoon, I resumed my journey, and traveled four miles to the residence of a cousin, John Barnherd, where I remained overnight.

The next morning I awoke to find a heavy snow falling, and by the time I was through breakfast, it was over a foot deep and falling fast. I was now twenty-five miles from home. No track had been made in the snow except by the postman's horse that had gone through pretty heavy walking, but I set off with good courage. The country was more settled than between Mexico and Constantia, and I occasionally found a sleigh track from one neighbor to another which aided me a little, but still the walking was very heavy, and I had hard work to make two miles per hour. About two o'clock P.M., I reached a more thickly settled road, and more traveled. When I had made about fifteen miles, and within ten miles of Rome, I came on a more traveled road. At this point I hoped for more relief than I found; more sleighs had passed, but the air was cold, the snow did not pack, and my foothold was like loose sand. About dark, when nine miles from home, I fell in with a man from Rome who was on foot, who relieved the loneliness of the balance of my journey. About nine o'clock P.M., I reached home after the most fatiguing walk I ever had.

The preceding account of traveling is of no importance except as to its historic value in showing the habits of men educated in a new country where they are under the necessity of helping themselves.

I closed my season's work about the 1st of January, 1820, after an employment of about nine months. I was instructed to charge for my services one dollar and twenty-five cents per day, with an allowance of fifty cents per day for board and traveling expenses; this I regarded as very satisfactory. Nothing was proposed for winter occupation, and I devoted the season to study and drawing. The most important book I had to study was the *Edinburgh Encyclopaedia.*[5] A copy of this work was owned by my uncle, John W. Bloomfield, to whom I have alluded. I had the full privilege of this, and found it afforded me a great source of information on the mechanical studies of canals, water works, bridges, bridge centers, and trusses. It was illustrated with numerous plans. In this way I passed the winter, not knowing what would open for the coming season. The canal was progressing east and west but I saw no opening for me. As the spring opened, I was relieved and found myself assigned to sundry duties on the middle section. I did not like this so well as the work on the new lines, which appeared more interesting as engineering work. I felt that some special favor had been extended to those who,

5. At this time (1820) there were no books in English on civil engineering in the United States. As late as 1838, D. H. Mahan, in the prefatory note in his text, *An Elementary Course of Civil Engineering* (New York: Wiley and Putnam, 1838), refers to "The Articles Bridge, Canals and Carpentary in the *Edinburgh Encylopaedia*" as one of the "best works to be studied."

from social position, had been preferred. But my prudence saved me from any expression of complaint and confirmed me in the faith that I must work my way by diligent devotion to the work I might receive, and that it was for me to do my work so well that my chief would not like to spare me.

The work assigned to me for this third season was rather of a miscellaneous character. There were several places on the middle section where the earth was of a very gravelly kind and proved to be quite leaky. Work was required to improve these places. It was found necessary to construct several small feeders to obtain sufficient water. I measured the whole section of about ninety-six miles and set up mile posts. These with other incidental duties occupied me until October. The discharge of these duties made it necessary for me to travel a good deal from place to place, and often I walked from ten to twenty-five miles a day, carrying a leveling instrument on my shoulder.

There was a very gravelly district between Limestone and Butternut Creeks, and being far distant from the feeders, it was not until midsummer that a boat was able to reach Syracuse. The public west of Syracuse were very impatient for navigation. In the western part of Onondaga County, wheat had been sold for thirty-one cents per bushel, and an improved method of reaching a market was greatly needed. This delay gave rise to much criticism as to the accuracy of the canal levels. Many intelligent men, among them the surveyor general of the state,[6] had expressed doubts of the skill of American engineers to establish especially long levels.[7] There was much discussion on this point in the public papers. As it was indispensable for a long

6. Simeon DeWitt was surveyor-general of New York State from 1784 to 1834. He was educated at Rutgers (then Queens College), and served as chief geographer of the Continental Army, 1780–81. DeWitt was very active in promoting the Erie Canal.

7. The Erie Canal commissioners had originally asked the English engineer, William Weston, to be chief engineer of the project. It was only after the aging Weston declined that, through the urgings of Commissioner Joseph Ellicott, an American, Benjamin Wright, was selected for the post. See William Chazanof, *Joseph Ellicott and the Holland Land Company* (Syracuse: Syracuse University Press, 1970), for a description of political intrigue surrounding the planning of the Erie Canal.

level (and this was sixty miles) that great accuracy as well as correct principle should be observed, the chief engineer directed Canvass White, in 1819, to rerun the level with great care.[8] In this he found small errors that were readily corrected, and it was believed to be correct. The circumstances of leakage, noted above, revived the controversy, but as the canal became tight, the water passed on and the accuracy of the level was established by the flow of water through its entire length. Considering that the level was run through heavy forests and dense swamps where in many places there was necessity of driving stakes to support the leveling instrument, the result grandly vindicated the ability of American engineers to establish any level.

It is hardly possible for the people of the present day to appreciate the extent of wilderness on the line of the Erie Canal at the time it was built. I have traveled from Rome to Canastota twenty miles, and after leaving Rome, I saw but one dwelling house, except the temporary erections by the contractors on the work.

When the canals were constructed, the State of New York contained about one and a quarter million people. A large part of the state was a wilderness, and the surplus earnings of its citizens were mostly absorbed in clearing the forests, draining the land, and in the erection of houses, offices, mills, and other improvements, imperatively called for by a people settling a new country. Of course the financial question was one of profound importance and, to a large portion of the people, appeared insuperable. I well remember the alarm of intelligent men that it would sink the state in irretrievable ruin. But little had been done in such works in this country at that time. The state sought aid of the general government. One of the commissioners, Joshua Forman Esq.[9] of Onondaga, said to me that after they had represented at

8. By modern standards of the leveling, the error would have to have been not more than about five *inches* for the entire sixty *miles*.

9. Joshua Forman was a prominent upstate New York lawyer and politician. He introduced a bill to the New York legislature for a canal between Lake Erie and the Hudson River in 1808. Forman worked closely with Wright in promoting the canal. Wright, for example, seconded Forman's resolution for the canal.

Washington [D.C.] their plan, and made request for aid, Mr. Jefferson [10] said to him, "We are trying to make a canal three miles long at this city, and we have not been able to obtain sufficient funds from individuals, the state governments, and the general government to complete it, and now you ask us to aid you in building a canal three hundred miles long through a wilderness. Preposterous!" Failing to obtain aid from other states or the general government, the State of New York entered single-handed on both great works, the Erie and Champlain Canals. The financial question, at that time a very serious one, called the attention of the best financiers of the state to its consideration. In this they devised a system and carried it out with such fiducial integrity that the state received a premium on five percent bonds of fifteen percent.[11]

As I have stated, my duties on the middle section occupied me until October. After this I was directed to aid the engineering work in the valley of the Mohawk below Little Falls, and remained there until the close of the season for such works. My small party was disbanded and I returned to Rome.

The eastern section was now under contract to and including Little Falls.

10. Thomas Jefferson was president from 1801 to 1809 during which time there was much controversy about the aid the federal government should give to specific local projects. The canal Jefferson refers to here was a small bypass canal. It was to run from Tiber Creek (near what is now the intersection of Constitution and 17th Street) through the center of the city along the north side of the Mall, turn south in front of the Capitol and enter the confluence of the Potomac River and Eastern Branch (Anacostia River) by what is today Fort McNair. This canal was built after Jefferson left office, but it was short lived. Today a few stone toll houses and a thoroughfare named Canal Street are all that remain of this project.

11. Jervis extracted this material from *A Memoir of American Engineering,* published in 1877 by the American Society of Civil Engineers as *Transaction* No. CXXXV. Jervis originally prepared it for his acceptance of honorary membership in the society. He was officially elevated to this position on December 2, 1868, and formally accepted it on June 16, 1868, at the society's national convention. It was on this occasion he presented the address, *A Memoir,* which was published nine years later. Hereafter, it will be referred to as *Memoir,* in Jervis' frequent excerpts from it.

The political power of the state was divided on the canal policy—those in favor were known as Clintonians,[12] and those opposed were mostly Democrats. These parties were a good deal mixed from the former parties of Federalists and Democrats. After the completion of the middle section, the canal policy came more decidedly into favor, and most of the Democratic party accepted it; as a result, large appropriations were made for the eastern and western sections, and the work was then much more vigorously prosecuted.

12. DeWitt Clinton was an active supporter of the Erie Canal project, and, according to Benjamin Wright, Clinton was its "master spirit." A lifelong politician, Clinton was a leader among the Federalists and in 1812 was the Peace party candidate opposing Madison's reelection. As governor of New York, with intermittent terms totaling seven years between 1818 and 1828, he was the arch-enemy of the Democrats who opposed any project he proposed, especially the Erie Canal.

III

SUPERINTENDING ENGINEER
ON THE ERIE CANAL

I REMAINED AT HOME during the winter [1819–20], having received the same compensation, one dollar and twenty-five cents per day, that I had the previous year. I improved the winter in the same studies that had occupied me the preceding winter, not knowing, but hoping my services would be called upon for some portion of the eastern section. When the spring opened and the arrangements for the season were made, I was informed by the chief engineer that I was assigned the position of resident engineer for the division extending from the Nose to opposite the village of Amsterdam, about seventeen miles.

This announcement was not only satisfactory, but very gratifying to me, and I entered on my new field with a determination that no effort of mine should be wanting to give satisfaction, and at the same time improve my knowledge of engineering.

The character of my new division for canal making was very different from that of the middle section, where hitherto I had been mostly occupied. Instead of the comparatively level surface of the latter, the line lay along the bold side hills, in many places coming down to the shore of the Mohawk River, requiring the canal works to be protected from the floods of the river by protective walls. The earthworks did not admit the same methods of computation as served on comparatively level ground, and consequently I had to learn a new method for this work. In this I was greatly aided by the principal assistant engineer, Canvass White.

There was also much more mechanical work, and on such as locks and dams I had no experience. The plans were all prepared in the office of the chief engineer [1] by his chief assistant, Mr. White, and I had only to study well what was required in the execution of the works. I was allowed two small parties, each having a leveler, and my duty was to attend alternately to each party. This made my duties more general as well as more responsible than they had previously been.

The canal line of my division had been located by Canvass White with the aid of George Young, a son of the late John Young of Whitesboro. Soon after I had entered on my charge, the line was placed under contract and work commenced.

Although the line was located, there was to some extent necessity for alterations, especially along the side hills. Sometimes this could not be fully known until the work was commenced, depending on the nature of the soil, as developed after excavation had made some progress. At the angles in the line, it was my duty to see that the proper curves were turned; this latter duty with that of placing the side stakes to guide the workmen was required to be commenced at once. I was required to prepare an estimate at the close of each month, showing the amount each contractor had earned during the month. On this estimate, funds were advanced to the contractor by the canal commissioner, Henry Seymour Esq.

A tour of the line was made each month during the season of work by the Chief engineer Judge Wright, his assistant (Canvass White), and Commissioner Seymour. This tour had the double purpose of making a general inspection of the works and of paying the monthly estimates. These heads of departments traveled in an open wagon, riding in part and walking over such portions of the line as was necessary, to inspect the works. These walks were sometimes continuous for five miles on a stretch. Occasionally, Mr. Seymour was accompanied by his son, a lad of twelve to fourteen years of age, who, I well recollect as a bright and pleasant youth, has subsequently occupied and still occupies

1. In later years, Jervis was to write that while Wright drew no plans for the Erie Canal, he was a "sagacious critic" of plans drawn by others and excelled them all in the vital element of practical judgment.

a very prominent position as a public man, under the cognomen of Horatio Seymour.[2]

Canvass White usually made an intervening visit in which he more deliberately attended to all questions of engineering duty. Holding strict ideas of discipline, I was very careful on those occasions to fully understand and strictly carry out all directions from my superiors, rarely making any question unless to explain a fact they had not fully appreciated. They rarely made complaint of my operations but often gave me encouraging words, implying satisfaction with the direction I had exercised. As time went on, and I had become more familiar with the wants of such works, I gradually began to criticize the plans, being careful to keep my own counsel until I had fully matured my views in every particular. I do not mean by this any radical change, but such incidental particulars as were developed in the prosecution of the works. The first suggestion I ventured to make which involved an engineering principle was in relation to the location of certain levels that appeared to me too low, more especially where streams were crossed at a level so near their beds as not to allow room for the floods to pass under the canal. In such cases, a system of dams and guard gates was adopted, by which a pool was made, constituting the best boat channel across the stream. These streams were turbulent and, I noticed in times of flood, brought down large quantities of gravel from the river bed above, filling in what was the boat channel. It appeared to me this would often be a serious obstruction to the navigation.

In a case that I had carefully considered, the natural level of the river bed was nearly the same as the bottom level of the canal. There was sufficient fall from the canal to the Mohawk River to carry off the floods, provided the gravel could be confined in its bed above. After much reflection, I came to the conclusion this gravel could be held in place by putting a dam across the stream at the head of the piers of the aqueduct, and thereby

2. Horatio Seymour was governor of New York twice, 1853–54 and 1863–64. A powerful influence in Democratic politics, he opposed Grant unsuccessfully in the presidential election of 1868.

retain the river bed above and prevent the gravel from washing down to choke up the passage under the canal. This dam forming a breast, and the water passing over and falling on a plank apron that extended across the stream between the abutments, kept the channel clear under the aqueduct. It had been designed to put a dam across this stream as before mentioned.

With hesitating caution, I ventured to present my views on this subject to my superiors and requested permission to be allowed to try the experiment on one of the streams we had to cross. My proposition induced my superiors to give it such attention as encouraged me, and after discussing the matter carefully, they granted permission to try the experiment. The result was completely satisfactory and gave me confidence to exercise judgment more freely than I had ventured to do previously. Subsequently the same plan was adopted for other streams. I am thus particular in this case, as it was the first suggestion on a point of hydraulic engineering I had ventured to present to the consideration of my superiors.[3]

About the first of January, after a season's service of about nine months, my parties were disbanded, and we went to winter quarters. No terms of compensation had been discussed, and I did not know until I came to settle my account with the chief assistant engineer (Mr. Canvass White), who was charged with this duty, what I was to receive. Mr. White directed me to charge at the rate of three dollars per day (about seventy-eight dollars per month). Board and incidental expenses were paid in addition. Our board cost, for the most part, one dollar and seventy-five cents per week. We lived mostly at private houses along the line of the canal, keeping with each an account of the number of meals we had, and counting twenty-one meals as equivalent to a week's board. This was a low rate for such irregular boarders. In some cases, the fare was quite indifferent, but in the most part it was comfortable and met the wants of men engaged in vigorous work.

3. An "experiment" such as this, without the theoretical framework but based on intuition, was typical of early American engineering. Another three decades would pass before men like James B. Francis would systematically perform hydraulic experiments based on theoretical considerations.

The compensation was fully satisfactory to me, and indicated that I would soon be able to pay the mortgage on my father's farm, which was then a matter I had seriously at heart. But as I was not prepared to fully make this appropriation at that time, I desired some method of investment that would give me interest until I should need the principal. As Mr. White was a man of much experience in affairs, I sought his advice on this matter of investment. He very frankly discussed the subject and gave me such advice as he considered prudent. Noting that I was prudent in saving my earnings, he entered into remarks on the philosophy of life, saying, a man, to have a respectable income, should save twenty thousand dollars, and have that safely invested at six percent per year. This, with a comfortable house in which to live, is sufficient for any man.

Mr. White was educated in the best society of Oneida County. I notice this incident as of historic value, showing the views of life that prevailed at that day.

In my service in 1818 and 1819 on the middle section of the canal, I frequently saw the commissioners, Myron Hawley [Holley], Samuel Young, and Ephraim Hart, but had no special business with them. My station was at that time too humble for that. In the spring of 1820, Henry Seymour, who had been one of the canal commissioners, and with whom I had frequent intercourse in relation to the duties I then had in charge, I became quite well acquainted with. This circumstance prepared me for a more familiar intercourse with him, when I entered on my charge in the valley of the Mohawk, he being the commissioner of that division in 1821.

The next year (1822) in my progress, I resumed the work on the division before mentioned. At the close of this year, my division was mostly completed and the parties disbanded, except myself. I was retained with a view to aid in the settlement of the accounts of contractors, which, with incidental duties, demanded my service for the entire year. Being economical in my habits, this year, at the previous rate of compensation, put my pecuniary affairs in a growing condition.

As before stated, the commissioner on this portion of the

canal was Henry Seymour; in the monthly tours over the canal line, he usually, as I have stated, accompanied the engineers. This brought my work into intimate review by him, as well as by the chief engineer and his chief assistant.

A very important business after the completion of a section was that of the settlement of the accounts of the contractors. At that day, more or less work in all contracts was omitted in the rates provided for in the specified items of contract and left to be adjusted by the chief engineer. In the discharge of this duty, Judge Wright directed his attention to such matters by his personal observations and discussions on his tours of inspection over the line of work. As these were rarely made a matter of record, when the time for final adjustment came, the chief engineer had necessarily to depend largely on the resident engineer as to many facts that would arise. Most contracts had more or less extra claims as it was by no means rare for ingenious contractors to swell these claims to large amounts.

During the progress of the work, I made a practice of taking notes, making measurements and calculations of such work as was likely to give rise to this class of claims. In this way I was not left to uncertain memory in cases of disagreement. This prepared me, as occasion would arise, to refer to my memoranda to sustain such view as I regarded correct and proper. By this course the chief engineer was able to dispose of such matters in a more just way, as well as in a higher confidence, than he could have done on any mere memory without such written memoranda. A gentleman stated to me a remark he heard from the chief engineer: "In any case of dispute on his division, he (Jervis) pulls out his papers, and that helps matters very much."

This closing up the accounts of my division brought me in very intimate relations with the canal commissioner and chief engineer. It gave me great pleasure to notice their unreserved approbation of my proceedings in the duties that had been committed to my charge. The canal commissioner took occasion to say to me that "my administration of the division had saved to the state thirty thousand dollars by my rigid attention to the records I was able to present." I may say with confidence, no approval of my proceedings ever came from even a remote hint on my part that it was looked for or expected; and I make these statements for the benefit of young men, that they may see the course that gives the highest satisfaction and is most sure to contribute to advancement. The whole is in a nut shell—a steady, resolute, discreet, and upright purpose is the basis of all worth, for any profession.

The principal assistant engineer had furnished blank books for each resident engineer, in which to enter a complete copy of his levels, notes, and computations of the several items of work on his division, and also the final estimate showing the cost of each section, with the view of depositing the information in detail in the canal department in Albany As soon as the business of settling contracts permitted, I set myself at work on this duty. It was a work of considerable labor to transcribe such an amount of figures and notes. When my book was completed, I handed it to the assistant engineer in the presence of his principal. They examined the book and expressed themselves satisfied with my work, remarking at the same time I was the only resident engineer that had complied with the instruction, and expressed doubts if any other had attempted it. They handed the book back to me, saying that if others complied, I might hand it in at another time; in the meantime I might keep the book for my own information. It was never called for, and I have the book still in my possession.

I have noticed the turn in public sentiment in regard to the canal policy. This turn brought the Democratic party into power, and the canal commissioners, being appointed by the legislature as their terms expired, these officers came from that party. In 1821, I think all except Mr. Hawley [Holley] of the acting commissioners were Democrats; but the honorary members of the board were Clintonian. The latter consisted of Governor Clinton and General Stephen Van Rensselaer,[4] who received no compen-

---

4. Stephen Van Rensselaer was a leading New York Federalist and militia general in the War of 1812. In 1824, while serving in the U.S. Congress, he founded one of the earliest technical schools in the United States, Rensselaer Polytechnic Institute.

sation. The duties of the latter were only at meetings of the board and for decision on important rules and regulations. The political antecedents of the commissioners were partly Federalist and partly Democratic.

The engineers, so far as I knew, were mostly of the Clintonian party, with similar antecedents to the commissioners. This disagreement in political views between the commissioners and engineers did not seem to mar the harmony of business. The chief engineer was appointed by the commissioners, and though he was of different politics, I never knew any effort to restrain him in the selection of his assistants; nor did I know of his making any appointment on political grounds. The commissioners and engineers worked harmoniously together and seemed to me as unitedly striving to secure the best results for the public interest. In selecting the contractors no notice was taken of their politics. At that day there was a creditable appreciation of public approval, and men had a high regard to the reputation they secured from their fellow citizens. I cannot say there was no political consideration in any respect; but so far as I had opportunity to observe, there was cordial and united operation on the part of the commissioners and engineers to conduct operations so as to secure the best interests of the canals. I had often opportunity to notice in their discussions of business the same earnest desire to reach the best possible result.

After I had entered on my duties on the eastern section of the canal, I had opportunity to become better acquainted with Colonel Samuel Young, who was the commissioner for the section between Schenectady and Albany. Also with Colonel William C. Bouck,[5] the acting commissioner on the western section of the canal. I had no special duties with these commissioners, but occasionally met them at their meetings on incidental business. At a subsequent period I was very closely associated with Commissioner Bouck, of whom I shall speak hereafter.

The eastern or Mohawk River section of the Erie Canal was

5. William C. Bouck was a professional politician and a leader in the antebellum Democratic party in New York. He was governor 1843–44 and held a minor federal position during Polk's administration.

nearly completed at the close of the year 1822. There was considerable work remaining to be done east of Schenectady. Above Schenectady there was little remaining, and most of this was between Schenectady and Amsterdam; a large, if not the principal part of this remaining work was work that had not been anticipated. This arose from peculiar circumstances which I will explain.

Schenectady is built upon a tongue of land that projects from the hill on the south side to the shore of the Mohawk River. Above this tongue of land, the interval lands of the Mohawk extend about three miles, and in the most part about half a mile wide, terminate at the upper end by the hill that strikes the river. The canal was locked down at the upper end of this interval land so as to carry a cutting or excavation through the central portion of this interval of one to six feet in depth. As this line neared the city, it curved off to the shore of the river, and so passed the tongue of land between the city and the Mohawk River. This made a cheap line to construct. This piece of canal was nearly constructed over the interval land, and some work was done along the river shore. This line had been opposed as being liable to damage from river floods, and as there was a local interest in the location, the matter was a good deal discussed among the citizens. One party favored the location as made, and another contended for a line through the center of the city, through the elevated tongue of land above described.

At that time it was considered the business of the city would center on the canal, and hence the local excitement. One party was led by Governor Yates and the other by R. Gevins, the proprietor and keeper of Gevins Hotel. The commissioners and engineers were guests at the hotel, and Mr. Gevins did not lose the opportunity of influencing, so far as he could, his guests on the question of location and was very active in looking after the route through the city which would most probably run near his hotel. At this time, a heavy flood occurred in the Mohawk River, rising over the banks of the canal, and gave great force to the objections that had been raised as to such dangers. This so impressed the engineers that they saw the necessity of some change, either by a new line or expensive guard banks to protect the canal in time of flood. This circumstance gave rise to the Gevins party, and though much work had been done, they succeeded in impressing on the engineers and commissioners to run a new line through the city, and finally in getting the canal authorities to vacate the river line and locate through the central portion of the city. The day after this was done, it so happened that Governor Yates and Mr. Gevins met at the halfway house, between Schenectady and Albany. Mr. Gevins said to me that Governor Yates called him out privately and gave him a severe reproof for the course he had taken, saying, "he (Gevins) was an uneasy yankee and could not be kept still." Mr. Gevins said he took the

The plan this measure gave rise to was to take up the lock at the upper end of this interval and raise the canal banks to correspond with the new level. The former excavation was not filled up, leaving the water in the canal six feet extra depth. The alluvial land was supposed to be watertight, but it was found after water was let in to be full of holes like pipe stems, made by the decay of aqueous roots, and gave a good deal of trouble to secure the banks against this kind of difficulty. In some cases long courses of sheet piling were put in, but the method adopted for the most part was to line the sides and bottom with sand from the hill. It was not until midsummer that this section of canal was so improved as to hold water for navigation.

In the opening of the season in 1823, I was retained by Commissioner Seymour to take charge of such work as was still to be done on the canal between the Minden Dam and the upper Mohawk Aqueduct, a section of fifty miles, excepting that Dr. Martineau,[6] the resident engineer on the division between Schenectady and Amsterdam, was retained to close up the accounts and work on his division, and continued in this service until about midsummer. The resident engineers of other portions of the section assigned to my care had closed their business and left.

It was now made my duty to organize parties of men to su-

6. John Martineau later served with Wright on the Chesapeake and Ohio Canal.

perintend the work of repairs and such incidental improvements as were found necessary to bring the section into use for navigation.

Mr. Seymour gave me general instructions, though the chief engineer, Judge Wright, was still acting, and I was authorized to confer with him in regard to the duties consigned to me. In the instructions given me by the chief engineer and commission, there was no reference to any political action but to employ men and provide materials with strict reference to the business wants of the work, so as to secure the greatest economy and efficiency of the arrangement.

It was essentially a new field of observation and experience to me. I had seen the water in the middle section of the canal, but this had few mechanical structures; the banks were mostly low, and for the most part composed of impervious earth, so there was little danger of breaks. The section committed to my supervision, as above stated, was quite different from the middle section. The mechanical structures were more varied and numerous—steep side hills and many places of rock cutting which required lining to make them watertight. The lining that had been used for this purpose was usually a clay loam. The earthwork over culverts, and at the joining with lock and aqueduct walls, was puddled. At that day, the puddling [7] was imperfectly done, and many serious leaks and breaks were the result. Clay is a naturally watertight earth, but if it be not thoroughly compacted, water will find its way through the spaces; when it does this, it gradually cuts away, enlarging the original leak. Clay never falls in to obstruct the leak. There is always difficulty in using clay to join with upright walls, as water will pass out between the clay and the wall, and especially when the frost of winter has thrown the clay so as to open a space or make the clay loose next to the wall. These circumstances caused much trouble in leaks and breaks at such places.

---

7. "Puddling" is a means for obtaining an earth embankment which is fairly watertight. Basically, it involves conveying water laden with soil to a pool or channel where its velocity is slowed and the soil particles settle out to form a blanket or lining.

It was the general practice at that time to repair such leaks by working in puddled earth, even when the water was in the canal. Repairs of this kind were often necessary to keep up navigation, and though the work could be much better done if the water were out, this could not be without arresting the trade of the canal. The latter consideration made it important to make such repairs without suspending navigation.

Finding the clay not satisfactory in the leaks that occurred, I made trial for this repair with fine gravel intermixed with sharp sand. I found this did not at the first application fully stop the leakage. But while the water percolated through the material, it did not sensibly carry it away, and the interstices being small, they were gradually filled up, and the work became tight, This material, when used in water, is much more effective than clay. The philosophy of this is obvious—the gravel does not pass off, though it allows some water to pass through it, and if any material is carried out, as it may be to some extent, the proximate material falls in and shuts up the opening; a few times of refilling makes the mass so impervious and solid as to retain all sediment and thus complete the repair.

In the autumn, when navigation had become quite lively, I had a leak at the head of a lock. The water passing down between the masonry and the earthwork and under the foundation forced up the plank in the chamber of the lock and discharged on the lower level. It was very important to maintain the navigation, especially as it occurred near the closing season, and I was unwilling to draw off the water if it could possibly be avoided. The leak baffled my efforts for several days, but was finally stopped by throwing in bundles of hemlock brush, following these at first with heavy gravel, and as the leak checked, by fine gravel. By this proceeding I so far controlled the leak that with watchful attention I was able to keep up navigation to the close of the season.

After the close of navigation I had the earth dug out at the head of the lock to the foundation. Then I extended the flooring of the foundation five feet above the head of the lock masonry, making with it a watertight connection so that water could not pass through without forcing its way under the earth and above the gravel. The theory of this is that while water will readily pass down a vertical joint between masonry and earth work, it will not find its way to any extent in a horizontal line under a superincumbent mass of earth. If the superincumbent earth is fine gravel and sharp sand, it may allow some water for a time to filter through, but not in quantity to do any essential injury, and will eventually become tight. After this repair I continued in charge of the section over a year, during which time not the slightest difficulty occurred with this lock, nor have I heard of any since I left the canal. This experience led me in all subsequent management of similar works to adopt as a rule an extension of a close planked foundation from four to six feet beyond the walls of masonry, and have found it to be a very effective method in such cases.

In the works of the Erie Canal it was a rule to allow culvert walls to rise to within one foor of the bottom level of the canal when the circumstances rendered it convenient to do so, and this often happened. This left one foot in depth of puddled earth over the top of the masonry. This depth of puddled earth did not prove sufficient for safety, and much trouble from leaks was experienced in this class of work. The masonry at that day was not very thoroughly made, which increased the difficulty. With good hydraulic masonry, and a covering of two feet of fine gravel and sharp sand, or by this depth of thoroughly puddled earth, the work may be considered safe, and this is as little as should be relied on. As a precaution in culvert works, the foundation should extend at least two feet beyond the walls of masonry, and two protective walls should be extended from the masonry not less than two feet on each side under each embankment to intercept any leakage that may occur in such parts. The same is necessary for culverts under the one embankment of a railway. It is to be observed that a culvert for a canal is much more exposed to damage by water than one for a railway or other road. The latter only requires the culvert stream to be guarded, while the former has this and also the water in the canal to expose its safety.

On the first opening of the Erie Canal for navigation, there was a fruitful source of embarassment arising from the method adopted for feeding one level of the canal from another. The method of operation was by setting the paddle gates so as to give the requisite flow of water from one level to the next below. During the day, when the lock tender could see what was wanted, he could easily adjust his gates. The difficulty occurred in the night, when for the want of a relief, the tender had to depend on his judgment as to the amount of opening necessary to give the supply while he was asleep. This required a skill and care for which there were no trained men or proper facilities. It often happened that either too much or too little water was passed in the night. This was more manifest on the lower levels, at times overflowing or endangering the banks, or leaving the water too low for navigation. Two evils were at time the result—one of the breaking of the banks and the other leaving too little water for navigation. In the latter case, the boatmen were not slow to manifest their dissatisfaction with the canal officials.

I recollect an occasion when I met the chief engineer at Schenectady, having myself just come in from the line; he inquired in his usual earnest manner of what I had to report of navigation then just commenced on this section. Among other things, I mentioned the difficulty with the lock tenders, as above mentioned. He remarked with much emphasis, "Those lock tenders have got no brains." A friend of the chief (Judge Raymond) had arrived at Schenectady with him, and was sitting by, listening to the conversation between myself and the chief engineer. At the remark of the chief above quoted, his friend very coolly inquired of the chief, "How much do you pay your lock tenders?" The chief replied, "Twelve dollars a month." Well, says his friend Raymond, "Do you expect much brains for twelve dollars per month?" The chief could only laugh at this very pertinent view of his friend.

It did not require a long experience to show that no skill of the lock tender was adequate to this duty unless he could have a night relief. A plan was soon adopted to pass the flow of water from one level to another by means of a self-regulating sluice in action at all times. This has become the general method for regulating the flow of water from one canal level to another.

There was a large portion of the eastern section of the canal that gave little trouble; but there were many points, some connected with mechanical structures and others in rocky and loose material of a porous nature, frequently causing breaches of more or less magnitude, and as before noticed, navigation to Schenectady was not fairly commenced until the autumn of 1823, and then with a light draft of water. It was a season of much activity, and the chief engineer (Judge Wright) and the canal commissioner (Henry Seymour) frequently passed over the work and gave me instructions for my guidance. They manifested great anxiety to establish regular navigation. This stimulated all in charge to the most vigorous effort. But with all we could do, there was great impatience on the part of friends and much severe complaining on the part of those who were opposed to the canal policy and management. It was impossible for those outside to appreciate the difficulties which had to be overcome. But there were not wanting intelligent men residing on the canal who, witnessing the activity manifested in conducting canal affairs and seeing in some measure the situation and difficulties, gave us encouraging words. These difficulties gradually grew less, and a pretty fair navigation was maintained through the autumn.

To refer to a more personal affair, I may state the close of this year put my finances in such a condition as to enable me to pay off the mortgage before alluded to on my father's farm and some other debts that pressed him, and now with a small balance began to save for myself. I have never paid out any money that gave me greater satisfaction than that above stated. To prepare for this object led me to the most careful economy in all personal expenditures that were consistent with proper regard for my situation. I had a good advisor in an aged uncle, John W. Bloomfield, before referred to, who always took a deep interest in my affairs. He was a man of sound judgment and of strict integrity.

In the spring of 1824, the canal was opened to Albany. I was continued as the superintending engineer on the same division I had the previous year.

It was important to the navigation that a higher level of water should be maintained in the canal than had previously been found practicable.

During this year the chief engineer, Judge Wright, had become so much occupied with new engagements that he spent but little time on the Erie Canal, which circumstance left me in closer relations with the canal commissioner, Henry Seymour. Mr. Seymour was my special advisor and attended to the settlement of my accounts.

As before observed, it was regarded important to sustain a greater depth of water for the navigation of the canal. This increased the pressure on the works, causing many new leaks and breaches where they had not before appeared. These, however, were not regarded in so serious a light as they had been, for by this time we had acquired considerable skill and dexterity in repairing them.

As these repairs had given a considerable degree of experience in their management, I will relate an instance.

About a month after the opening of the canal, in the spring of 1824, Mr. Seymour met me at Schenectady where we had a consultation on the affairs of the canal. At this time a breach had occurred at a point about sixteen miles below Schenectady, and twelve miles beyond my district. Mr. Seymour had just come up from Albany and manifested much anxiety about that break, and did not seem to know how it was progressing in repair. To relieve this anxiety, he proposed that I should go down and examine the situation and report the same to him at Utica, whither he was on his way. I said to him I did not like to do that, as my neighboring superintendent might be jealous of my meddling on his territory. There was the best of feeling between myself and my neighbor, but such an act on my part was liable to be misunderstood. To this Mr. Seymour replied that he did not wish interference at all with their operations, and was only desirous to obtain from me a report of the facts of the case and learn the probable time when the repairs would be completed. The next morning Mr. Seymour left for Utica, and I mounted my horse and rode for the break. I met an assistant superintendent who did not seem well pleased with my presence; but I did not at all meddle with his operations, and after surveying the situation of things, I set my face back for Schenectady. The day was very hot, and a ride of thirty-two miles in the saddle brought me in with a pretty sharp headache. I threw myself on my bed with a view to obtaining rest and relief. In half an hour a knock at my door called me up. On opening the door I saw one of my overseers, who informed me that a break had occurred about four miles above Schenectady. This I thought a sharp close of my survey of the affairs of others.

Of course there was work on hand, and I ordered my horse and started for the break. The sun was about half an hour high, and I reached the break in time to take a thorough survey. I found the break had been caused by the breaking in of the covering stone of a culvert, and had taken out the bank of the canal for a space of about forty feet in length and some twelve feet in depth. There was no time to repair the culvert; the bank must be repaired and navigation restored, leaving the masonry of the culvert to a future time. To do this, it was necessary to remove the culvert stone that lay under the bank. This was a work of considerable labor and had to be done before the breach in the bank could be safely filled in. After setting the mode of operation, it then being dark we went to supper. The men of the overseer were on hand, and after supper a party of men went to work clearing out the break. We found it a troublesome work to remove the culvert stone, as some of them were quite heavy and the earth soft. By this time I had forgotten my headache, and remained all night preparing the break for filling. It was necessary to clear the stone, so as to guard against leakage, and to slope up the ends of the breach, so as to admit teams to pass in readily with material to fill the break.

We got all in readiness soon after the sunrise the next morning. Men had been sent out during the night to engage teams to haul in the filling. We then went to breakfast. Teams began to come in soon after breakfast, and by eight o'clock the work of filling was in active progress. In order to secure dispatch in such cases, there was placed in the breach four men who stood ready, two at each end of the wagon to lift the wagon-plank with the greatest promptness so as to dispatch the team and allow another to come in its place. This is the most expeditious method I have known to repair a break. It makes a very lively time and calls out a brisk activity with the men. The level of the canal where this break occurred was about four miles long. By one o'clock P.M., we had the break so far filled that I ordered the water to be let in, being confident we could keep the work up to meet the rise of the water. At six o'clock P.M., we had sufficient depth of water, and a packet boat passed over the level, and before the next morning, loaded boats moved regularly. That night I slept soundly, and the next morning felt none the worse for my previous night's labor.

The method of filling a break as above related is very effective against leakage and allows water to be let on immediately; moreover, is the most expeditious as well as the most economical that I have known. As to compacting of the filling, this is secured by the equal and regular distribution of the material and the compacting by the horses' feet and the wheels of the wagon.

In my report to Mr. Seymour, I had to include work on my own section; and I could say loaded boats passed by break and were still detained at the one I had visited at his request.

I continued as the superintending engineer until March, 1825. My service for 1823 was not in full charge, as some service was performed by the resident engineer east of Amsterdam. For the year 1824 I had full charge for my entire section, and all accounts for labor and material passed through my hands. At that day there was much that was peculiar in the expense of maintaining the canal—in repairing breaches, watching the banks for early discovery of any indication of a breach, and the gravelling of the towpath. The latter had not been provided for in the original construction.

The actual cost of maintaining this section of fifty miles for one year, including lock tenders, and all expenses except that for the collection of tolls, was at the rate of six hundred dollars per mile. I suppose there was no section of the canal of equal length that had more difficulties in the way of expense of maintenance than this section at that time. And this should have been less expensive in subsequent years. At this rate the entire canal (363 miles) should have been maintained at a cost of $201,800 per year. It must be remembered, however, that the canal commis-sioners and chief engineer were not then made by election or general ticket. No political advisors were at that time able to mar the business affairs of the canal.

Whatever may have been the views of men high in official station, it was not regarded proper to interfere with the economical conduct of business on the canal. In all my intercourse for seven years, no intimation was given me to look to the right hand or the left for any motive but the strict interest of the canal. When superintending, commissioner Seymour did not in the least give direction or even intimation as to whom I should employ in any department of work. I selected the man I wanted with strict reference to the ability I supposed he had for the work to be done.

The chief engineer, Judge Wright, in 1823 had many calls for his services on canal enterprises in other states, and these so occupied his time that I had only occasional visits from him. Mr. Canvass White, the principal assistant engineer, left the state service for other works early in 1823. My principal advisor in 1824 was Henry Seymour Esq., canal commissioner. With him I had the same pleasant and cordial intercourse I had enjoyed with the chief engineer and his principal assistant.

There was during 1824 considerable call for engineers on works started in several places. Though I highly appreciated the experience I had obtained in the two years I was charged with the superintendence of the canal under navigation, I was desirous of engaging in new works. With this purpose, I informed the canal commissioner of my intention to close up my engagement on the state canal. After announcing my intention to Mr. Seymour, he urged me to continue my superintendence, saying he was satisfied that engineers were best qualified for superintending and maintaining the navigation of the canal, and that he would employ such as were willing to perform this duty. I had no complaint to make as to compensation, or as to the agreeableness of my relations with the commissioner, who promised to make my salary as liberal as circumstances would justify. Still, as an engineer seeking new fields of occupation, I naturally looked to new enterprises.

I had no special work in view, except as both Judge Wright and Canvass White had intimated prospects for employment on new works if I concluded to leave my superintendency on the Erie Canal. I therefore felt a reasonable confidence in obtaining an engagement.

After several conferences with Mr. Seymour, at which he seemed to think I would recede, I finally said to him, I must leave. He appeared unwilling to consent, holding the impression that I would abandon my purpose, expressing evident disappointment at my conclusion. A day or two after my course was decided, Samuel Young Esq., an associate commissioner, said to me, "Mr. Seymour appears as though he had lost a friend in your leaving the service." It would be vain for me to disguise the pleasure I felt in this evidence of appreciation from one who held a trust of high position in the confidence of the public, and with whom my limited business relations had been very intimate, cordial, and pleasant.

IV

DELAWARE AND HUDSON CANAL

AFTER seven years' employment in the engineering department of the Erie Canal, I closed my services early in the month of March, 1825.

I then took a steamboat for the city of New York, which was my first passage on a steamboat.

Soon after my arrival in New York, I had an interview with the Honorable Benjamin Wright, the former chief engineer of the Erie Canal. A short time previous to this interview, Judge Wright had entered into an engagement with the Delaware & Hudson Canal Company as chief engineer of their proposed canal.[1] At the insistence of Judge Wright, I had a conference with the committee of the board of directors. This conference resulted in my engaging to serve as principal assistant engineer under Judge Wright as chief. At that time, Judge Wright held the position of chief or consulting engineer for several other works, and under his advice, it was understood that I should organize the engineering force and superintend the general duties of the service. This was to me in great measure a new responsibility.

While I had the benefit of the advice of the chief, it developed on the assistant to examine the route, make surveys and establish the location of the canal, and also to prepare the plans and specifications of the various structures required for the canal. In this duty, the discussion of all particulars and especially of all difficulties was as a matter of course referred for the decision of the chief.

---

1. Wright served as chief engineer of the Delaware and Hudson Canal from May, 1823, to March, 1827. The Delaware and Hudson Company is still an operating corporation. It was originally chartered in 1323.

Considering I had only left my engagement on the Erie Canal early in the month of March, and that my new engagement was consummated on the 12th of the same month, not much time was lost in the change of service.

The preliminary survey [May–November, 1823] and report of the projected canal had been made the previous year [actually in January, 1824] by John L. Sullivan Esq.[2] of Massachusetts, under instructions from Judge Wright, who was employed by the projectors of the enterprise, the Messrs. Wurtz of Philadelphia.[3] On this report a charter had been obtained for the incorporation of a company from the legislatures of Pennsylvania and New York, giving the necessary authority to construct the canal. Under these laws a company was organized in the city of New York. Phillip Hone of New York was the first president, and John Bolton, treasurer.[4]

Immediately after my engagement, I received instructions from the chief engineer to make an examination of the country through which the projected work was to be made, and to follow this with a report on the characteristics of the enterprise. With these instructions, and the reports and maps made the previous year, I set off for the field of operations. The next morning I arrived at Kingston, Ulster County, the eastern terminus of the

2. John Langdon Sullivan was the son of Governor James Sullivan of Massachusetts. With degrees from both Harvard and Yale, he left the business world to study English and French canals. He succeeded Loammi Baldwin as the engineer of the famed Middlesex Canal (Boston to Lowell) which his father had helped found. President Monroe appointed him to the Board of Internal Improvements in 1824. In addition to the studies of the Delaware and Hudson Canal, Sullivan also made a survey of the Chesapeake and Ohio Canal. Later, he left engineering and practiced medicine, after obtaining his medical degree at the age of sixty.

3. The Wurtz brothers, William and Maurice, were the originators of the Delaware and Hudson Canal scheme. Their younger brother John, who was a congressman at the time (1825–27), later joined the D&H Company and succeeded Bolton as president in 1831, serving until 1858. Another brother, Charles Stewart, was not active with the company.

4. Phillip Hone, a prominent New York businessman and politician, became mayor of New York in 1825 and president of the D&H Company on March 11 of the same year. John Bolton was elected company treasurer on March 12, 1825, and succeeded Hone as president of the company on January 21, 1826. He served until 1831, and most of the original construction was during his tenure.

proposed improvement. I had with me John B. Mills, who had been Mr. Sullivan's assistant in the previous survey.

Mr. Mills and myself traveled on horseback from Kingston, through the Mamakeating Valley sixty miles to the Delaware River, and up to this river about three miles to Saw-mill rift. From this point we saw no roads, and it was impracticable to travel in the saddle in the valley of the river. We therefore sent our horses round by way of Milford to Mount Moriah on the Lackawaxen River, a distance of about thirty miles. We then proceeded on foot with our examination of the Delaware River section.

We found a more rough country for a canal along the lower ten miles of the Rondout Valley than I had been accustomed to see. Though there were rather serious obstacles on the route between the Hudson and Delaware Rivers, which could only be partially understood by such an examination, the country generally had a reasonably fair look for a canal.

On passing up the Delaware Valley, the severe aspect of the bold steep shores of the river, and its rapid current, looked very unfavorable as compared with anything I had before seen as a route for a canal. Our first proceeding after leaving Saw-mill rift was to pass over hills that formed the shore of the river at Butlers Falls, a distance of about one mile. The path was about three hundred feet above the river. At the time of this examination, the ground was covered with about one inch of snow which had fallen the previous night and made the walking on the steep side-lying ground rather difficult, requiring us to seize the bushes and limbs of trees for support. As we began to descend the hill above the falls, we had a fine view of the river below. The bold rocks rising nearly perpendicular from fifty to two hundred feet above the river, the turbulent action of the water at its base, with the general gloom as heightened by the snow and the wild surroundings of the scene made an impression on my mind that fifty years have not eradicated. It certainly presented a very unfavorable situation for a canal, and did not fail to impress the difficulties of the enterprise.

The same precipitous shore of the Delaware continued with modifications to the mouth of the Lackwaxen River, a distance of seventeen miles, though no other as severe as that at Butlers Falls. At several points we were compelled to leave the shore of the river and follow foot paths over the rocky ledges. At Meeteck Falls, in passing along one of these paths, my foot was caught in a bear trap that had been set in the path and was so covered with leaves as to be hidden from view. Fortunately, the jaws of the trap caught the heel of my boot, and no harm resulted. The trap was large enough to have broken my leg, had I struck it in the right place. I mention this as indicating the wild condition of this route at that time. The water in the river was at a good rafting pitch, rather high, and the occasional passage of a lumber raft was essentially all the stir that evidenced the civilization of this district.

The Lackawaxen Valley, for about fifteen miles from the mouth of the river, was of a bold character but less severe than the portion of the Delaware Valley above described. The remaining ten miles to the junction of the Dyberry was comparatively moderate. We arrived at the mouth of the Dyberry, one hundred and five miles, in about eight days from Kingston.

We proceeded up the valley of the west branch of the Lackawanna, seven miles to Keenes Pond, the head proposed for the canal. This valley from the mouth of the Dybury was less precipitous than that of the Lackawanna below. What is now the flourishing village of Honesdale at the junction of the Dyberry and West Branch was then a dense forest, and in fact there was but a very small portion of the land settled after we left the Neversink Valley.

In his report before referred to, Mr. Sullivan had terminated the navigation portion of the improvement at Keenes Pond and proposed a railroad thence to the coal mines in the valley of the Lackawanna at Carbondale. Mr. Sullivan suggested in his report that he was of the opinion further examination would show the practicability of carrying the canal on from Keenes Pond to the coal mines. I saw nothing to warrant this suggestion, and supposed it had its origin in the views of that day on the superior economy of canal transportation.

After visiting the mines at Carbondale, we retraced our steps, giving special attention to such prominent objects on the route as demanded more careful attention than we were able to give on our first examination. Mr. Mills afforded me the aid he was prepared to give, from his survey of the previous year, by which I was better able to give attention to the more difficult characteristics of the enterprise.

Mr. Sullivan's plan was to make a considerable portion, about three-fifths of the work, between the Hudson and the Delaware Rivers, through the Mamakeating Valley, a regular canal, and the remaining two-fifths a slack-water navigation.[5] After reaching the valley of the Delaware, his plan was mostly by locks and dams forming pools or slack-water by navigation. This last was a feature of the enterprise which required to be decided at the outset, before the work of construction could be commenced. This was an important matter in my examination, and I had made it the subject of special attention. It did not appear to me that the characters of the rivers were favorable for Mr. Sullivan's plan so far as regarded the slack-water navigation. The rapid fall in the rivers, and their great rise in time of floods, requiring large expense in dams, guard locks, and guard banks to protect the navigation of short pools, led me to the opinion that an independent canal would be at least as cheap, and at the same time afford a better navigation. I therefore reported in favor of an independent canal as most suitable for the enterprise.

As before stated, Mr. Sullivan proposed the termination of the canal at Keenes Pond, seven miles from the junction of the Dyberry (now Honesdale), and as a railway was proposed, and was the only kind of improvement between the coal mines and Keenes Pond that appeared to me as justified by the formation of the country, I approved of the report in this respect.

The country between the forks of the Dyberry and Keenes Pond was not difficult for a canal, except that it required a large number of locks for the distance—about thirty locks of ten feet

5. A "slack-water navigation" is a system for modifying a river by placing dams below rapids so as to flood them out, permitting cargo barges to pass over them safely. Locks or inclines are provided at the dams to permit passage of the boats.

each in seven miles. So large a number of locks would greatly increase the cost of the canal, and this was not the only objection, as they would materially increase the current expense of navigation. On the ground of the cost of construction, and the increased cost of navigating a canal with so large a number of locks, I was strongly of the opinion that the railway should be extended down to the forks of the Dyberry.

I made my report in accordance with the remarks above made, and a few weeks later, the chief engineer, Judge Wright, and myself made a second examination of the route of the proposed improvement between Kingston on the Hudson and the coal mines at Carbondale. At that time I suppose Judge Wright was about sixty years of age [actually, he was fifty-five]. The country between Kingston and the Delaware was not difficult to explain. The water in the Delaware was not so high as on my former visit, and we could pass along the edge of the river in many places, either on foot or by the aid of a small boat or skiff which accompanied us, that we could not pass on foot at my former visit.

In following our route up the valley of the Delaware and Lackawaxen Rivers from what is now Port Jervis to Mount Moriah, it was necessary to make this distance of thirty miles on foot, excepting the occasional relief afforded at some points by the small boat that accompanied us. We took the journey as easy as we could, the chief sleeping in his cloak which he had wisely taken with him, in the log houses where we found quarters, and some of which were pretty rough. The object of the trip made it necessary to give considerable attention to the local features of the route, requiring three days to reach Mount Moriah on the Lackawaxen. Though considerably fatigued, the chief bore this pedestrian portion of our examination very well. We then proceeded to the completion of our examination, to Carbondale.

The chief engineer, after completing this examination, approved in the main the report I had made, and it was decided the work should be an independent canal (except about two miles of Eddy Pond near tide water) from tide water at Eddyville to the junction at the Dyberry (now Honesdale), a distance of one hundred and five miles. This left the question of extending the canal from Honesdale to Keenes Pond for future consideration.

Regarding the formation of the country as presenting many difficulties for such a work, with one hundred and ten locks on one hundred and five miles of canal,[6] and to a large extent the want of suitable materials for masonry, and considering the limited means at command for such works at that day, it was a bold project, and evinced on the part of the board of directors a hardy spirit of enterprise. The main object of this canal was to furnish means of supplying fuel to the city of New York and the towns on the Hudson River. At that day the city of New York received mineral oil from Richmond, Virginia, and from Nova Scotia and England, the latter mostly brought as ballast.[7] The balance of fuel for the city was wood. No anthracite coal was at that time used in New York. It was therefore very important to establish by cheap transportation a channel to the anthracite mines in Pennsylvania. At that time anthracite coal was supplied to Philadelphia from the Mauch Chunk Mines, and floated down the Lehigh River on arks.

The Delaware and Hudson Canal enterprise was brought before the public by Maurice Wurtz Esq., and his brothers of Philadelphia. They owned coal mines in the valley of the Lackawanna River and originated this enterprise to open a market for their coal. They obtained the survey and report of [Wright and] Mr. Sullivan before noticed, and also the corporate power from the legislatures of Pennsylvania and New York, under which the company was organized. The capital stock was mostly subscribed in New York, and that city was made the business center of the company's operations.

In connection with the canal, it was necessary in order to reach the coal to construct a railway over the Moosic Mountains,

6. This was about one lock per mile. For comparison, the Erie Canal had one lock per *four* miles (eighty-eight locks in three hundred and sixty-three miles). These one hundred and ten locks added about eight to ten hours to the travel time.

7. "Ballast" is heavy bulk material carried by a ship in compartments below its water line to provide stability in rolling seas. "Mineral oil" was coal.

as before noticed, as a part of the enterprise. On the whole it was a very formidable undertaking in any view, but especially at the time it was entered upon. It could not promise any return for the funds invested until several years of persevering work. It was necessary to meet the influence of river coal interests that soon sprang up and made war on the Delaware and Hudson in the fear it would interfere with their own projects. This at times caused considerable agitation. But the board of directors pursued their onward course with great firmness, being deeply impressed with the importance of the work to the city of New York and the Hudson Valley. The board of directors having settled the general plan of the work, surveys were immediately commenced for the definite location of the canal. John B. Mills Esq., the engineer before mentioned, was placed in charge of the first locating party, commencing his work at tide water on the Hudson. The line was located to Mamakeating on the summit, a distance of forty miles from the Hudson, and put under contract. I examined the line personally, traveling over it on foot and portions of it several times. After the line was prepared and maps and profiles made, the chief engineer made examination with me and gave his approval of the location of line and plans and specifications of work.

The duty of location involved much care and labor. At the same time, the work of preparing plans and specifications of the several mechanical structures devolved on the principal assistant engineer [Jervis]. My headquarters were at Kingston, but a large part of my work was done at Mamakeating, now Wurtsboro. At the time I am speaking of, the village of Kingston presented a number of the ruined walls of buildings that had been burned by British troops in the war of the Revolution.[8] I well recollect the anecdote of Mr. Elmendorf, who ran a small wagon to convey passengers from the village to the steamboat landing. An Englishman had occasion to take passage, and rallied the old

8. The British under Sir Henry Clinton had been moving north along the Hudson to join with Burgoyne. A marauding force of his army learned that the New York legislature was meeting at Kingston and sacked the city on October 7, 1777.

man, inquiring why they did not drive the British away, and prevent the burning of the village; to which the old man innocently replied, there were only women and children there—the men had all gone to take Burgoyne at Saratoga.

The board of directors appointed Maurice Wurtz Esq. their general agent to make contracts and pay the estimates for the work.

It was thought advisable to have a public ceremony for breaking ground, and accordingly there was a large gathering at Mamakeating to witness the actual operation of President Hone in taking out the first earth removed for the canal [July 13, 1825]. A dinner was provided, and John Suydam Esq. made an address in addition to that of the president. I could not see the value of all this parade for so small an amount of work, and so expressed myself to one of the directors present. I well recollect the decided expression of his countenance as he looked at me, saying, "What would this world be good for if there was no good eating." The director was Mr. Hasbrouck of Rondout.[9] The celebration of breaking ground being over, I went to work, and in the month of August, 1825, work on the line was begun in earnest.

The whole line between the Hudson and Delaware Rivers was put under contract the same year. Towards the close of this year a company of people assembled at a place then known as Carpenters Point. It was a fine table of land, commanding a good view of the broad valley of the Delaware below. It was the point where the canal line left the valley of the Neversink, and proceeded up the Delaware. This assembly gave the name of Port Jervis to this place. I was pleased at the wholly unexpected compliment, though at the time I had no idea of the prominence to which the place has grown.

After the agent, Mr. Wurtz, came on the line and entered on his duties, the board gave us an open two-horse wagon, and a pair of horses for our traveling accommodation. These, with the aid of a horse and buggy I afterwards found necessary for my

9. Abraham Joseph Hasbrouck, banker and politician, was an incorporator of the D&H Company. He served in the House of Representatives in 1813 as a Clintonian Democrat.

occasional use, enabled us to take the chief engineer when he had occasion to travel with us on the route of the works. It was common for the agent and myself to travel on horseback, in which case the single buggy carried our baggage and the provisions we found it convenient to have with us for dinners, and enabled us to dine at such convenient places as the business required.

As the business of the line progressed, other assistant or resident engineers were employed. The most experienced of these, after Mr. Mills, was James S. McEntee Esq., who had been an assistant of mine on the Erie Canal.[10] Most of the others had little practical experience of such duties, and from the want of experienced assistants, I had no small difficulty in conducting operations. But as time went on, I was able to improve this service by men who developed sagacity and showed themselves able to take higher places than they were first called to fill. Among these I may mention James Archbald [11] and Horatio Allen.[12] Mr. Allen had been sent to me from the Chesapeake and Delaware Canal [13]

10. McEntee apparently admired Jervis so much that he named his son Jervis McEntee. Little is known about the father, but the son became a nationally known artist in the second half of the nineteenth century.

11. James Archbald came to America in 1805 from Ayrshire, Scotland. He became a contractor on the Erie Canal where he first met Jervis. About 1825, Jervis recruited him for the D&H Canal, and he had a long term of service, later taking charge of the famous gravity railroad. His ability and his friendship with Jervis probably led to a vice-presidency of the Michigan Southern and Northern Indiana Railroad in the 1850's. When Jervis left the MS&NI in 1858, Archbald also left and became president of the Delaware, Lackawanna and Western Railroad. Archbald, Pennsylvania, is named after him.

12. Horatio Allen was perhaps the most famous of Jervis' younger protégés. The son of a mathematics professor at Union College in New York, Allen graduated from Columbia in 1823 and, after a year studying law, joined Wright's engineering staff on the Chesapeake and Delaware Canal in 1824. He moved with Wright to the Delaware and Hudson Canal where he first met Jervis. From 1829 to 1834, Allen was chief engineer of the Charleston and Hamburg (South Carolina) Railroad. It was the longest in the world when it was completed in 1833. After three years abroad, Allen rejoined Jervis on the Croton Aqueduct from 1838 to 1842. He then went into the iron business, but also maintained a small consulting practice. From 1871 to 1873, Allen served as the fifth president of the American Society of Civil Englineers.

13. The Chesapeake and Delaware Canal was first proposed in 1769, but not opened until 1829. Only fourteen miles long, the principal obstacles to its completion were legal rather than technical.

by Judge Wright. He was a man of good education and readily entered into the duties required. Mr. Archbald was a man of great sagacity of mind, and proved a most valuable appendage to the department. Afterwards I obtained the services of Porteous R. Root,[14] who had several years of experience on the western section of the Erie Canal, and John T. Clark,[15] who had experience on the eastern section of the Erie Canal.

Without further detail I may say the surveys and contracts were steadily carried forward to the termination of the line of canal at Honesdale.

A dam was constructed across the Delaware River a short distance below the mouth of the Lackawaxen River. This had two objects, one to feed the canal, and the other to form a pool to pass the boats from the canal on the Delaware to the canal on the Lackawaxen side. It was necessary to make a sluice in this dam to pass rafts over it. The dam was about six hundred feet long, and the rafting sluice about one hundred feet. The sluice was cut down three feet and filled with flush plank supported by brackets to hold the water during navigation, removed during the rafting season. The apron of the sluice did not prove satisfactory to the raftsmen, and it became necessary to improve it before the spring season for rafting. This was in the month of March, and the water from the melting snow was very cold. In order to do the work, it was necessary to put up the flush plank and brackets to turn the water from the apron. At this time the water flowed lightly over the dam, and of course on the sluice it was three feet deep and running with a strong current. The task of putting in the flush plank under this state of water was extremely difficult. A party of carpenters and laborers had been provided for this work. A service that would require a good deal of nerve to overcome the difficulties of the case, I committed the charge of the work to three resident engineers—James Archbald, John T. Clark, and Russell F. Lord.[16] The carpenters halted and

14. Root later became chief engineer of the Black River Canal in New York.

15. Clark went into railroad engineering after leaving the D&H Company. He rose to leadership of the Hartford and New Haven Railroad.

16. R. F. Lord should not be confused with R. L. Lord, an original "manager" of the company. R. F. Lord served as engineer in charge of the canal

said it was impossible. The engineers found it necessary to take the lead and go forward into the water. Such was the force of the current, they were only able to support themselves by reaching out and boring holes in the apron timbers, into which they drove a long pin and this pin was their foothold for that reach of their work. And so step by step they carried the work forward, the carpenters following. They began immediately after breakfast, and the flushing was completed at two o'clock in the afternoon. The engineers drank no liquor, but the carpenters thought whiskey indispensable; the latter gave out before the work was done, and left the engineers alone to complete it.

I relate the above in justice to those engineers whose energy, skill, and moral nerve accomplished a work that few men would have undertaken. After their work, they changed their wet clothes for dry, and reported they felt no harm from the severe exposure. Their minds supported their bodies. I could only regard them with profound admiration and here record their doings to their honor.

The chief engineer, Judge Wright, being engrossed with other works,[17] resigned his position in 1827, and I was appointed to the head of the engineering department.

THE GRAVITY RAILROAD

It had been decided to terminate the canal at Honesdale (and construct a railway from that place to the mines at Carbondale), one hundred and five miles from Kingston. The railway was sixteen miles in length, and over an elevation of about nine hundred feet to the summit on Moosic Mountain, and about nine hundred and fifty feet descent thence to the mines. At that day, such work

---

from 1830 to 1863. It was during this time that the four famous Roebling aqueducts were built.

17. Wright was then engaged with Joseph Totten, Canvass White, and General Simon Bernard on the engineering board of the Chesapeake and Delaware Canal across the northern neck of the Delmarva Peninsula. Since 1824 he had also been a consultant on the James River and Kanawha Canal in Virginia. In 1828 he became chief engineer of the Chespeake and Ohio Canal in Maryland and consultant on the Blackstone Canal in Rhode Island.

involved very serious engineering questions. The rise in elevation was by no means regular. On either side of the mountain it had the largest portion within two to three miles. It was obvious from all the experience of that day that the only economical method of operation was by stationary machinery. Railways in mining districts in Europe had been constructed and operated by stationary power. No general traffic had then been accomplished by railways.

In July, 1827, I made a visit to Boston with the single object of ascertaining what I could learn from the Quincy Railway, then just put in operation.[18] I mention this to show the narrow range of experience at that time on such subjects in this country.

The works on the canal were so far advanced in 1827, that my attention was called to the special consideration of the railway portion of the improvement. As before stated, it had been decided to construct a railway from Honesdale to the mines at Carbondale, a distance of sixteen miles. The ascent as before stated from the mines to the summit of the Moosic Mountain was about nine hundred and fifty feet. In a distance of three miles, this ascent was provided for in five inclined planes worked by steam power. After a summit of one and one half miles along the side of the mountain, the line began to descend toward Honesdale. The first mile had a descent of about three hundred and fifty feet. This descent was put into one inclined plane.

18. Perhaps the first railway in the United States was a short, three-quarter-mile tramway at the quarry of Thomas Leiper near Philadelphia which was built c. 1809–10. However, the Quincy, Massachusetts, Railway of 1826 was probably the first "engineered" railway. Other claimants to the "first" title were a short brick haulway on Beacon Hill in Boston (c. 1795), a power mill haulway near Falling Creek, Virginia (1811), and a nine-mile gravity railroad built at Mauch Chunk, Pennsylvania, by Josiah White and Erskine Hazard in 1827. The Quincy Railroad was carefully planned by the pioneer American engineer, Gridley Bryant. It had many of the elements of a modern railroad; tracks (iron-covered wood) on sleepers, guard rails, a turntable, and four-wheeled trucks which were sometimes used in a swiveling configuration, but no steam locomotives. The truck device was the subject of the famous patent litigation, *Ross Winans v. New York and Erie Railroad,* c. 1840. Bryant was to approach Jervis by letter (July 27, 1839) and offer to build the High Bridge Aqueduct out of the famous Quincy Granite. His offer was not accepted, probably because of its cost.

There were two other descending planes, and the intermediate railway was on a descent, varying from twenty-five to forty feet to the mile.

The great question was to devise the best plan of machinery to work the planes. Most of the machinery that had been put into operation for such purposes was of English origin. They were sufficiently described in books that I was able to procure. I was not satisfied with these plans. They appeared to me too cumbersome in their works and did not secure the convenience of operation necessary to the best economy. I aimed to dispense with the large drums they used and to substitute in their place a sheave wheel.[19] In this it was necessary to provide sufficient hold on the sheave to prevent the prepondering load from slipping the chain in the sheave. To do this I had spurs let into the sheave to hold the chain. The spurs held the chain and prevented its slipping in the sheave, but in some way this caused the chain to break, and the method had to be abandoned and hempen ropes substituted. To guard the rope from slipping, I had a second sheave so placed as to carry the rope back to a second groove in the main sheave, giving it double the hold in the sheave. This was completely successful.

The next important feature was to provide something better, more convenient, and more reliable as a means of regulating the loaded trains in passing down the descending planes. In this case the prepondering power was large, and required a strong check to control it. The friction break had been depended on for this object. It occurred to me that the atmosphere could be made to do this work, and if so, it would be more safe and much more convenient. I examined this feature with much care, made numerous experiments on the resistance of air, and devised an apparatus which I called a pneumatic convoy, by which the passage of trains was secured in a very perfect manner. I took the precaution of putting a friction brake into the machinery so as to pro-

19. "Large drums," in this context, are cylinders around which cable is wound; "sheave wheels" are wheels grooved to fit cables such as in a pulley. Jervis recommended an "endless" cable running through sheaves as the hoisting system in place of a drum which would wind up the cable each time hoistings were required.

vide for any failure in the pneumatic convoy. As the first train passed over the angle of the plane, I watched the action of the convoy with great interest and had the pleasure of witnessing its perfect success. The train passed down the plane with the most perfect regularity.

A road graded at forty feet to the mile extended about six miles from the foot of the second descending plane to the head of the third plane. The latter was subsequently taken up by my very able successor, Mr. James Archbald, and the forty-feet grade continued to the head of the canal at Honesdale. The empty cars were then hauled up an inclined plane by steam power, and from this a line of rail ran on a grade of forty feet to the mile into the bottom of the valley and then by another plane to a high point from which the empty cars ran on a graded rail as far as the descent permitted; thus alternately, by inclined planes worked by steam power, and a graded rail that allowed the cars to descend by gravitation, ten miles of railway were worked by stationary power and gravitation, which was a decided improvement.

The board of directors submitted my report to Professor Renwick of Columbia College.[20] Mr. Renwick went to Carbondale to see me, and I was anxious to learn his opinion on my report. It was a very full report, going freely into detail and the discussion of the subject. He said there was some error in my calculation of water power that I had recommended for one of the ascending planes. He did not state what it was, but I went immediately to my room and examined my report, and to my chagrin I discovered the error. I then fell back to steam as I had proposed for the other ascending planes. Mr. Renwick made a

20. James Renwick was born in England but came to New York as a child. A scholastic prodigy, he graduated first in his class from Columbia when only seventeen years old. Although frequently called upon as an engineering consultant, Renwick was principally a professor of science at Columbia from 1812 to 1853. He was an arbiter on many technical decisions; the one mentioned here by Jervis was typical. His son, James, Jr., worked with Jervis on the famous Croton holding reservoir located where the New York Public Library now stands. Renwick the younger became a famous architect, his works including St. Patrick's Cathedral in New York City and the Smithsonian Institution in Washington, D.C.

full report to the board. Excepting this error he found nothing to criticize, and as a whole spoke favorably of my report and expressed confidence in my ability to successfully conduct the work.

My report was also submitted to Judge Wright, my predecessor. Judge Wright did not go into a general criticism; but advised the use of horse power instead of steam for the ascending planes. In other respects he made no decided criticism. After these reports had been submitted as above for counsel, the board of directors instructed me to go forward with the work, leaving me without any specific advice to proceed on the general plan I had recommended in my report. Of course they furnished me copies of the report of counsel. My plan was therefore carried out, with the exception of the substitution of steam power for water power on one of the ascending planes. At a subsequent date Mr. Archbald constructed a railway on the plan of using no moving power for the Pennsylvania Coal Company. This railway is about forty miles long and by inclined planes worked by stationary power and graded road to allow descent by gravitation; it is worked in both directions in the same way. This plan can only be adopted in a country that has high elevation of proximate hills above the valley, in which the railway is built.

As an instance of the service of those days I state as follows: On an occasion I left New York in the afternoon by steamboat and arrived at Newburgh about nine o'clock in the evening. The fatigue I had in New York gave me a severe headache. At Newburgh I was to take a stage to a point about ten miles from Honesdale. I was to be called up at two o'clock A.M. I felt very little like taking the stage so early, but went to bed having given orders to call me in time. When called for the stage, I felt very poorly. However, I got up and entered the stage. It was late in October, the night dark and the roads very muddy. The stage was full of passengers, and we were nine hours in making twenty-two miles. By this time my headache was gone and we took breakfast. After breakfast we had a long hill to rise, and myself and one or two others thought to walk up the hill that was in our way. It was about a mile to the top. Reaching the top of the hill considerably in advance of the stage we walked on; when the stage came up, we were at the foot of another still longer hill, and so we concluded to go on up that hill. In this we had gained so much on the stage that one other passenger beside myself concluded to keep on. After traveling about nine miles in the second stretch the stage again overhauled us, and we got in. About six o'clock in the evening we arrived at the next place for eating. After supper we started off, and about five o'clock A.M. we arrived at the next eating place, and by ten o'clock in the morning I arrived at the end of the stage route, having traveled sixty-six miles in thirty-two hours. I then took a one-horse wagon and drove ten miles to Honesdale, the end of my journey, arriving in good condition for business. I mention this to show how different traveling was in those days as compared with the present.

The works of the canal and railway were completed in the autumn of 1829, and a few boats loaded with coal were transported to tide water on the Hudson.

I recommended to the board of directors as superintending engineers Mr. James Archbald for the railway and mines and Mr. Russel F. Lord for the canal. They had risen from rodmen on this work and manifested excellent executive ability for such service. The board confirmed my nominations, and they were continued about thirty years (Mr. Lord until his death) in charge, and to their ability and integrity the ultimate success of the enterprise was largely indebted. Mr. James Archbald had an excellent engineering mind and great practical sagacity and was eminently upright in purpose. Mr. Russell F. Lord was a man of good executive ability and indefatigable in industry. I employed Mr. John H. McAlpine to superintend the construction of machinery. He was an accomplished machinist and discharged his duties in a very satisfactory manner, and was for many years continued as superintendent of machinery. Mr. McAlpine introduced to me his son, William J. McAlpine,[21] and requested a place for

21. William Jarvis McAlpine was among the most distinguished American civil engineers of the later nineteenth century. In addition to his work with Jervis on the D&H Canal and the Erie Canal enlargement (1836), he was chief

him in one of the engineering parties. The son was then about sixteen years of age, a bright, active, and pleasant boy. Advancing from station to station he was for many years in my employ; manifesting capacity, industry, and fidelity, he was one of my most esteemed assistants. His subsequent career and standing is before the world as among the most eminent engineers of the country.

I have alluded to the hardy character of this enterprise. It was early evident the cost would materially exceed the original estimate. Of this the board of directors were advised. They, however, had so high an impression of the value of the improvement, in which they were sustained by the chief engineer, that they firmly resolved to go forward. Regarding the serious character of the impediments to the work, the engineering department pursued their duties with unremitting devotion to its interest. Those who had some experience, and those who entered on the work as learners, with very little exception, prosecuted their duties with energy, industry, economy, and fidelity.

The department had but one office for construction business, and for this only rent was paid. It was at the headquarters in Kingston; the office had two small rooms, and was rented for about eighty dollars per year. The offices of resident engineers were the rooms in which they lodged or some room furnished by their landlord as a perquisite for the patronage the engineers brought to his house. Not much furniture was provided, except some plain tables and drawing boards. No one appeared to have any desire to look after personal conveniences or gratification, their minds finding all the amusement they needed in the intense interest they had in the success of the work. The movement of the resident engineers with their parties along the line of work was almost invariably on foot. The company had few carriage bills to pay.

There was no special discipline for the purpose of securing a

---

engineer or principal consultant on the Brooklyn drydock, the Erie railroad, the "Eads" Bridge, Manchester Ship Canal (England), and many other major projects. McAlpine served as state engineer of New York and was president of the American Society of Civil Engineers in 1869.

vigorous prosecution of the work on the part of the engineering department. My own mind was naturally devoted to the success of the enterprise, and absorbed in the responsibile duties I had undertaken. My intercourse with my assistants was frank and cordial, discussing with them freely all matters of business connected with the work, listening patiently to their suggestions, and finding ample material for discussion in the numerous exigencies of the work.

In the course of operations there were many difficulties that developed from time to time, but they were overcome, and the final success accomplished. The scarcity of funds induced methods and plans that would not have been adopted had it not been for the necessity of great economy. No doubt this enabled us to get the work into operation, but more ample means and better work would have been true econmy. As we had not the funds, we regarded it wise to do the best we could with what we had. Rarely I think, has any company been more faithfully served.

Most of the men who were employed in the engineering department were disbanded as their works were completed, except as I have mentioned Mr. Archbald and Mr. Lord who remained as superintending engineers. Some of them abandoned engineering and followed other vocations. Of these were Mr. James S. McEntee of Rondout and Mr. Henry Farnam,[22] since deceased, of Port Jervis. The former is still living, a highly respectable and independent citizen. Others continued in the profession and became more or less eminent for their abilities. Mr. John T. Clark and Mr. William J. McAlpine were both, at different times, state engineers for the State of New York.

In the spring of 1830, I resigned the position of chief engineer of the Delaware and Hudson Canal Company, and its affairs were conducted by the two superintending engineers before mentioned. I, however, engaged to give occasional attention for the next year to aid in the adjustment of unsettled business and give my advice on the general management of their affairs. At

22. This engineer is not to be confused with Henry Farnam, also a civil engineer, after whom Farnam Hall at Yale is named. At this time, the Yale Farnam was chief engineer of the Farmington Canal in Connecticut.

the expiration of this last engagement, I resigned all charges of their works, having been employed over six years.

As I have stated, the chains for the inclined planes were a failure, and ropes were substituted. The ropes and all the machinery worked well. As I shall notice hereafter, the locomotives did not succeed, and horse power was substituted for the sections where locomotives had been intended. I have observed the horse power was afterwards abandoned and the work done by stationary engines. After the railway had been worked by horse power several years on the parts that had been designed for locomotives, stationary power was introduced.

When the improvement was put in operation there was considerable difficulty on the canal portion, and navigation was often embarrassed by leaks and breaks in the canal. In 1831 a pretty fair business was done, and I think about one hundred thousand tons of coal was transported.[23] The finances of the company were limited, and for a time the sale of coal did not meet expectation. But year by year the difficulties were overcome, and the sale of coal increased. The company had a severe struggle for several years, and their stock was low on the market. By perseverance they gradually improved in the conduct of navigation and in the sale of coal, and in a few years reached a high state of prosperity. I think much of their their prosperity was due to the able and faithful engineers (Archbald and Lord) in their respective departments, and to the indomitable energy of John and Maurice Wurtz, aided by an intelligent board of directors, among whom were I. L. Platt, I. Hawley, General Ewing,[24] and others who gave much personal attention to the affairs of the company, and had great satisfaction in bringing the institution into a flourishing condition. In the supply of fuel to the city of New York and the other towns in proximity, it was a most valuable improvement. Subsequently the canal was enlarged and the tonnage of boats increased.

23. Actually, about fifty thousand tons were transported in 1831, but in 1829 only seven thousand tons were shipped.

24. Isaac L. Platt served on the board from 1834 to 1852, Ira Hawley from 1842 to 1852. No "Ewing" is found among the managers (directors) of the D&H Company.

V

EARLY LOCOMOTIVES

THERE have been several erroneous statements made through the public papers in relation to the failure of the locomotive engineers provided for the railway of the Delaware and Hudson Canal Co. Heretofore I have not thought it worthwhile to notice the error of the assertion that the railway was in fault. The subsequent history had demonstrated the railway was abundantly adequate to sustain all the service I had intended it should. The fact that in a few years its traffic had risen from one to two hundred thousand tons per year was good evidence the swaying and shaking complained of was mainly a matter of imagination.

An article in the New York *Times* of August 9, 1877, calls up this subject again and is calculated to perpetuate the erroneous impression as to several facts of the case and has induced me to make some corrections. The *Times* states; "Although the locomotive as a locomotive was a perfect success, the railroad was not calculated to stand its use." That the engine was too heavy for the railway, there was no doubt. The railway was a cheap structure; that is it was of such materials as the country afforded, and was carefully calculated to bear a certain weight, about one ton on a wheel. All which it sustained under a larger coal traffic, until the timbers failed from natural decay.

I made a very detailed report on this railway, carefully examining the elements of strength, and did not intend to have a load beyond its strength. The question of locomotive engines was at that day little understood. I brought it forward in that report and satisfied myself it was a power that could be used with economy on some sections of the railway and so recommended, and the Board of Directors decided to authorize the experiment.

As I have stated, the rail was designed to carry one ton on a wheel. This was for the cars, and when I applied it to the locomotive, I adhered to the same weight, and in order to obtain more power on one section of the road, I proposed six wheels for the locomotive, all bearing one ton on each wheel. Further reflection led me to doubt whether the working of six wheels in one frame would do well on our very curved railway.[1] Very little experience of any kind was then had and especially of six-wheeled locomotives. In this view, on the 23rd of April 1828, soon after his arrival in England, I wrote to the agent[2] who had been appointed by the company to purchase the locomotive, stating my doubts as above and requesting him to examine the subject, and if it did not appear clear to adopt a six-wheel engine, we must be content with a four-wheel. In such event, I said to him, I thought our railway would carry an engine of five and one half tons on four wheels. That I regarded this to be the extreme, and he might obtain one engine of this weight, but to have all others of lighter weight or of five to five and one quarter tons. The five and one half tons would be one and three eighths tons on a wheel, and the five tons would be one and one quarter on a wheel.[3]

1. Jervis' doubt was well-founded. Later he was the first to develop the "bogie truck" to overcome this problem. The bogie truck is a frame supported by two pairs of small wheels. The frame itself is attached by a swivel to another frame which supports the front section of a locomotive's boiler. This arrangement steadies and guides the locomotive on curved sections of track. For this reason, it is also called a "pilot truck." Before the development of the truck, the size, hence power, of a locomotive was limited. George Stephenson's first six-wheeler was his "Experiment," completed in 1828.

2. Horatio Allen was only twenty-six years old at this time. Jervis had written Allen on January 11, 1828, specifying maximum weights of locomotives (six to seven tons for six-wheeled and five and a half tons for four-wheeled). He also indicated they should be capable of attaining five miles per hour and that the "chimney" (stack) should not be greater than ten feet tall.

3. Altogether four locomotives were purchased for the D&H Company by Allen; they were probably named "Lion," "America," "Delaware," and "Hudson." All four were delivered to New York City during 1829, but only the "Lion" is known to have been assembled and operated. The "America" cannot be

The Stourbridge Lion, the only one of the locomotives that was set up on the railway, weighed seven tons exclusive of fuel and water, and with its proper complement of fuel, and water in the boiler, about eight tons, or fifty percent beyond the extreme limit I had given. Now in the light of locomotive experience of the present day, it will probably be said, this was an insignificant size for a locomotive. But it must be remembered, the locomotive "Rocket" built by George Stephenson [4] about the same time for the great contest, and which won the triumph on the Liverpool and Manchester railway in October, 1829, weighed only four and one quarter tons; one of its competitors weighed four and three quarter tons, and all the rest on four wheels were of less weight than four tons. This trial [5] was made two years after my report for the use of locomotives for the Delaware and Hudson Canal Company had been made. So it is clear I was fully up to the times as to the weight of locomotives. Subsequent experience with locomotives on railways has shown my engine would have had double the power I had calculated.

When I understood from the agent what the weights were likely to be, I felt much solicitude on the question, though at that time I did not fully understand what the weight would be. The agent was to procure the locomotives under my direction. As I have said, the question of locomotives was very imperfectly understood at that day, and I gave the agent no instruction as to the plan of the locomotives, leaving that to his judgment after examining the results of experience in England. But in relation to weight I was particular, lest it should be greater than the rail would carry.

---

traced beyond its transshipment point at Eddyville, New York (July 16, 1829); the other two may have been destroyed, unassembled, in a warehouse fire in Rondout, New York, sometime after September 22, 1829. Some correspondence indicates that the "America" was also called the "Pride of Newcastle" after its city of manufacture.

4. George Stephenson was the most famous English engineer of the early nineteenth century. He was not only a principal developer of the steam locomotive, but also of total railroad systems.

5. The so-called Rainhill Trials were held about two months after the road tests of the "Lion."

The experiment was a great loss to the company and to me a very serious disappointment. It was clear the agent had not carried out my instructions. In such a case, it would seem to have been a dictate of ordinary prudence to have kept within my limit, stated as it was regarded the extreme that would be safe and rather that he should have erred in less, than greater weight, especially as his original instructions were a weight of one ton on a wheel. If as a measure of caution, the weight of the engines had been five instead of five and one half, it could not have been under the circumstances a matter of complaint. As the road had some surplus strength over the calculated load, such an engine would have realized the expectation. And as subsequent experience has shown, an engine of four tons would do more work than at that time was estimated for one of six tons. I then estimated the tractile power [6] as one to seventeen and it is now well settled it may be regraded as one to six, and if prudence instead of boldness had ruled our agent, our hopes of locomotive power would have been realized and the loss to the company and disappointment to myself averted.

I did not feel disposed to make severe criticism on the agent; it was a day in which the subject of locomotives was in its infancy and I considered the builders had misled him.[7] I did not think it reasonable to put the failure of the agent on the railway instead of the locomotive; and withal to give great credit to the man [Horatio Allen] who ran the engine on trial,[8] for his courage

6. "Tractive force" would be more accurate: the ratio of the load on the drive wheels as compared to the weight which could be pulled by the locomotive without the wheels slipping. If the load on the driver is ten tons, and the tractive force is one-sixth, a pull of ten-sixths tons (one and two-thirds) can be developed by the engine. Later work indicated this ratio is between one-fourth and one-fifth.

7. Robert Stephenson and Company of Newcastle, England, built the "America" ("Pride of Newcastle"); Foster, Rastrick and Company of Stourbridge, England, built the other three locomotives.

8. The first trial run was on August 8, 1829, and the second on September 9. After these two unsuccessful trials, the "Lion" was put in storage for twenty years. From about 1849 to 1869, the boiler alone was placed in service at Carbondale, Pennsylvania. The Smithsonian Institution received the boiler in 1889 and scattered parts since that time.

in a case in which no special courage would have been called for if the agent had executed his mission according to instructions.

The New York *Times,* in the article referred to, states; "After quitting the docks, the track turned abruptly to the west by a curve of most threatening radius, and crossed the Lackawaxen River over a slender hemlock trestle, nearly one hundred feet high." As to this, there was no trestle work at this place nor at any other on the whole railway, of anywhere near one hundred feet in height. At the bridge referred to, it was about twenty-five feet in its greatest height. Of course I do not remember the exact height, but it did not vary materially from twenty-five feet. It therefore seems the great height was assumed to give romance to the article. This note of the *Times* is particular to put the failure on the railway instead of on the locomotive and the agent who procured it. As I have stated elsewhere the railway has fully sustained its design.

The *Times* further states of this trial trip "dashed away from the village [Honesdale] around the curve, and over the shaking and swaying trestle at a rate of speed estimated at fifteen miles per hour." In this it is sufficient to notice this same "trestle work" and this "threatening curve" did afterwards, bear a coal tonnage of between one hundred thousand and two hundred thousand tons per year, and no one engaged on it seemed to suppose there was any great heroism in conducting it. As to the point of failure, the Resident Engineer, Mr. Archbald, examined the effect of the locomotive and reported to me that the weight of the engine so impressed the light iron plate into the wood that he was of the opinion it would not do to rely on it as a means of transportation. When this report was received I had not supposed it would be necessary to abandon the use of the locomotives. I then went over the track that had been occupied by the "Stourbridge Lion" with Mr. Archbald and examined it with him. This confirmed Mr. Archbald's report. We did not see any failure in the trestle that was regarded important, nor anything to cause doubt as to the strength of the railway, to sustain the traffic it was designed to provide for and this has been abundantly demonstrated in its subsequent use, having borne the transit of the first twenty years of about five million tons.

Now it is a little remarkable that with so much danger staring him in the face, the trial trip runner should have "dashed away" at a speed nearly four times as great as the business requisites of the locomotive called for. But notwithstanding the great courage manifested, he did not seem to forget some degree of caution, and so moved the engine "slowly up and down" and then after feeling the strength of the rails, he "dashed away," probably having satisfied himself the danger was imaginary, and the apprehensions of the crowd a very natural result of his operations on those who had not the least idea of such operations.

As a reason given for this courageous experiment, the *Times* gives what purports to be the inducement, as follows: "The runner saw the great danger at a glance, but feeling that locomotive power on railroads was destined to become universal in years to come, the pride of possessing the distinction of having been the man to direct the energies of the first one on the continent overpowered his sense of danger and he declared that he would make the trip, let the consequences be what they might." It might be inferred from this that the runner had lost sight of the object he was employed to secure, namely, the trial of a locomotive engine that would serve a useful purpose on this railway, and had resigned himself to a dangerous feat for the purpose of securing personal ambition, to have lost sight of his proper responsibility in making trial of locomotives designed to secure the business and success of his employers.

Having myself all preliminary responsibility of procuring the locomotives, I might say, I was the first to introduce on "this continent" a locomotive engine. Whatever honor might be attached to this (and I think it would be equal the honor of acting as the first man to run it), I did not think it worth while to assert. The commercial object had been defeated, and the Delaware and Hudson Canal Company were involved in a heavy loss, and I was sorely disappointed. Under the circumstances I was in no spirit to assume the honor of introducing the first locomotive on "this continent." What was painful to me, in the loss of the commercial value of this part of the enterprise, appears to have had no importance with the man who ran it on the trial trip. When it is considered this loss arose from the failure of the locomotive, we can hardly account for his ambition in so striking off to a field altogether out of the mission he was employed to fill.

There are some other historical items in the *Times* article; but as they do not affect this railway, nor the "Stourbridge Lion," I shall not attempt to correct them.

As I have stated, I was not disposed to criticize the agent in this matter, and should not now have alluded to it had it not been for the attempt to throw the failure on the railway, which was not responsible as I have shown. The agent was left freely to his own discretion, except as to the weight of locomotives, which latter was material to the harmony of the project. The matter of this trial had passed away from public view, and I certainly never attempted to keep it alive and was not a little surprised to learn that after more than forty years it should have been thought worthwhile to bring it out in a way that was calculated to keep out of sight the proper responsibility in the affair. I therefore make this record from the necessity that has been thrown upon me to vindicate the history of this early railway. I do not do it from any hostility to the man who ran the engine, the agent. In regard to the agent, I always supposed he was misled by the maker of the locomotive, and had shown more confidence in the strength of the rail than I had; and as the subject was in its infancy, I was disposed to consider the failure as attributable to the want of knowledge on the subject, though I could not have been more particular than I had been on the question of weight.

THE LOCOMOTIVE TRUCK

Some notice of this truck is presented in the above extract, but as the invention has been claimed for Mr. Horatio Allen, it appears proper for me to give a more full account of the invention.

In a review of the treatise [9] of William Brown on the "American Locomotive," as published on the 9th of December 1871, in the *Rail Road Gazette,* this claim for Mr. Allen is made. I am unable to say that he makes this claim himself, as the criticism appears as editorial, and I have no right to say it was dictated on any other than editorial responsibility. From the article referred to I make the following extracts.

Mr. Brown's book also contains an engraving of the double truck engine [10] built for the South Carolina Railroad by Horatio Allen, of which we published an engraving in the *Gazette* of March 4, 1871. These engines were not only the first eight-wheeled locomotives which were ever built (if the editor will look in the first edition of Wood on railways,[11] he will find a plan and description of such a double truck engine), but they were the first to which the truck or bogie system was applied in any form, for which too much credit cannot be awarded to Mr. Allen.

In this connection we think it is but fair to state there is some discrepancy between a claim made by Mr. John B. Jervis, in a letter published in the book we are reviewing, and some of the testimony given in the Winans celebrated car case.[12] In the letter referred to, Mr. Jervis says: "I was the inventor and put in successful operation the locomotive truck." At the time of the trial referred to, he gave the following testimony. "I invented a new plan of frame, with a bearing carriage,[13] for a locomotive

9. *History of the First Locomotives in America* (New York: D. Appleton and Company, 1871).

10. Called the "South Carolina," it was built at the West Point Foundry in 1831 under specifications supplied by Allen. It was delivered to Charleston in January, 1832, and after five months of vexatious breakdowns, it was modified in such a way as to provide satisfactory service; orders for four similar locomotives were authorized.

11. Nicholas Wood, *A Practical Treatise on Railroads and Interior Communications in General* (London: Longman, Orme, Brown, Green & Longman, 1825). Wood was a close friend of both George and Robert Stephenson and a pioneer in the development of steam locomotives.

12. Based upon a patent granted him in October, 1834, Ross Winans, a Baltimore inventor and locomotive builder, brought suit against a series of railroads for their unauthorized use of eight-wheeled cars with movable "trucks." The ensuing legal process extended up to the U.S. Supreme Court which, in 1858, decided in favor of the railroads.

13. The "Experiment," with two large-diameter drive wheels and a four-wheel swivel truck ("bogie" or "bearing carriage") in front, was the fifth (and

engine, in the latter part of the year 1831, for the use of the Mohawk and Hudson railroad, which was constructed and put on the road in the season of 1832. The trucks in this engine worked perfectly; but the boiler, being intended for anthracite coal, did not do well, and another boiler was made for it. Soon after this a second engine, with the same plan of wheels and bearing frame, was made and put in operation on the Schenectady and Saratoga Railroad early in the year 1833.[14] The engine had six wheels; on one pair, the drivers rested in the usual way on one end of the frame of the engine, the other end of the engine rested on the frame of a four-wheeled car or truck, so arranged that by means of a center pin passing through the transom beam, the upper frame on which the engine rested could follow the guide of the lower frame, without necessarily being parallel with it. . . . By this means a long frame for an engine could be and is supported near its end, which provides for the most steady motion of the machine, and by the separate car or truck to guide, it passes on curves with all the facility of a short-geared car."

From this testimony it will be seen that the "new plan of frame" or "bearing carriage" referred to, was invented in the latter part of the year 1831, and put on the road in 1832.

Now from the testimony of both Mr. Allen and Mr. C. E. Detmold [15] given on the same trial, it is clearly shown that the plans for Mr. Allen's double truck engines were made in the winter of 1830 and 1831, and that the first engine built on these plans was constructed at the West Point Foundry in 1831, received at

---

last) locomotive built by the West Point Foundry in New York City. This machine is said to have attained a speed in excess of sixty miles per hour in trial runs.

14. The "Davy Crockett" was built to Jervis' specifications by Robert Stephenson and Company (though Taylem and Company was also mentioned as the builder) during the winter of 1832–33 and shipped to the United States on April 6, 1833. The first paying trip was on July 2, 1833, but the first field trials may have been held in May on tracks of the Mohawk and Hudson by Asa Whitney.

15. Charles E. Detmold, a German-born engineer, later became a successful businessman. In his early career, he drafted the plans for the first locomotive built by the Kembles at the West Point Foundry, the "Best Friend of Charleston" which was designed to Allen's specifications. Its weight was five tons or one and a quarter ton per wheel—just as Jervis had specified for the English locomotives in 1828. Detmold had also designed the horse-powered treadmill car for the Baltimore and Ohio Railroad in 1828.

Charleston in January 1832, and put in operation February, 1832.

The truck or as our English friends call it, "the bogie," was designed and applied to locomotives by Mr. Allen nearly a year before Mr. Jervis used it. It is true that the latter gentleman employed it in a somewhat different way, and used only one truck on his engines instead of two, as Mr. Allen did, but the latter and not Mr. Jervis, as he claims, was the first to devise and apply the truck to locomotives. It is true that the peculiar application which Mr. Jervis adopted is the one that has come into general use in this country, and it is only since Mr. Farlie, with an energy almost unparalleled, has resuscitated the use of the *double* truck, that the plan which Mr. Allen adopted has come into use.

I made a reply to the above criticism, so far as related to my truck. My reply was published in the *Gazette*. As it discusses the important points in the above criticism, I transcribe it in this connection.

Rome, N.Y.
13th Dec. 1871

To the Editor of the Railroad Gazette:

In your issue of the 9th instant, page 376, you criticize Brown's history of locomotives. As my name is introduced, I hope you will allow me space for explanation.

In the discussion of any matter, it is important to understand clearly what the subject is. You state that the truck, or as our English friends call it "the bogie," was designed and applied to locomotives by Mr. Allen nearly a year before Mr. Jervis used it. The question is, what kind of a truck did Mr. Allen use? And what kind did Mr. Jervis use?

In the summer of 1830, Mr. Horatio Allen was the chief engineer of the South Carolina Railroad and I was at the same time chief engineer of the Mohawk and Hudson, now a portion of the New York Central Railroad. Mr. Allen was allowed to spend his summer at the North. He occupied the privilege by spending nearly his whole vacation with me at Albany, as he also did the following summer. We were both very much interested to ascertain some method by which the weight of the engine could be spread on more wheels, work more favorably, and thus save our weak railroads from an action they could not safely support. On

Miniature of John Bloom-
field Jervis, probably
painted at the time of the
Croton project.

Engraving of John Bloom-
field Jervis, made near the
end of his life.

The birthplace of John B. Jervis, 434 Park Avenue, Huntington, Long Island, as it appears today. Now known as the Powell House, it is a locally prominent example of early American architecture. *Courtesy of Huntington Historical Society*

Jervis' two-span stone arch aqueduct over Rondout Creek near High Falls, New York, was built in 1826 for the Delaware and Hudson Canal Company. It was replaced by Roebling's cable-suspended aqueduct in 1849 and demolished in 1956. *Courtesy of Manville B. Wakefield*

The Delaware and Hudson Canal lock just south of the Mongaup aqueduct about six miles north of Port Jervis, New York. Both lock and aqueduct were built by Jervis late in 1826. Both sets of gates are open in this photograph so that the fall of water over the upper sill can be seen. *Courtesy of Manville B. Wakefield*

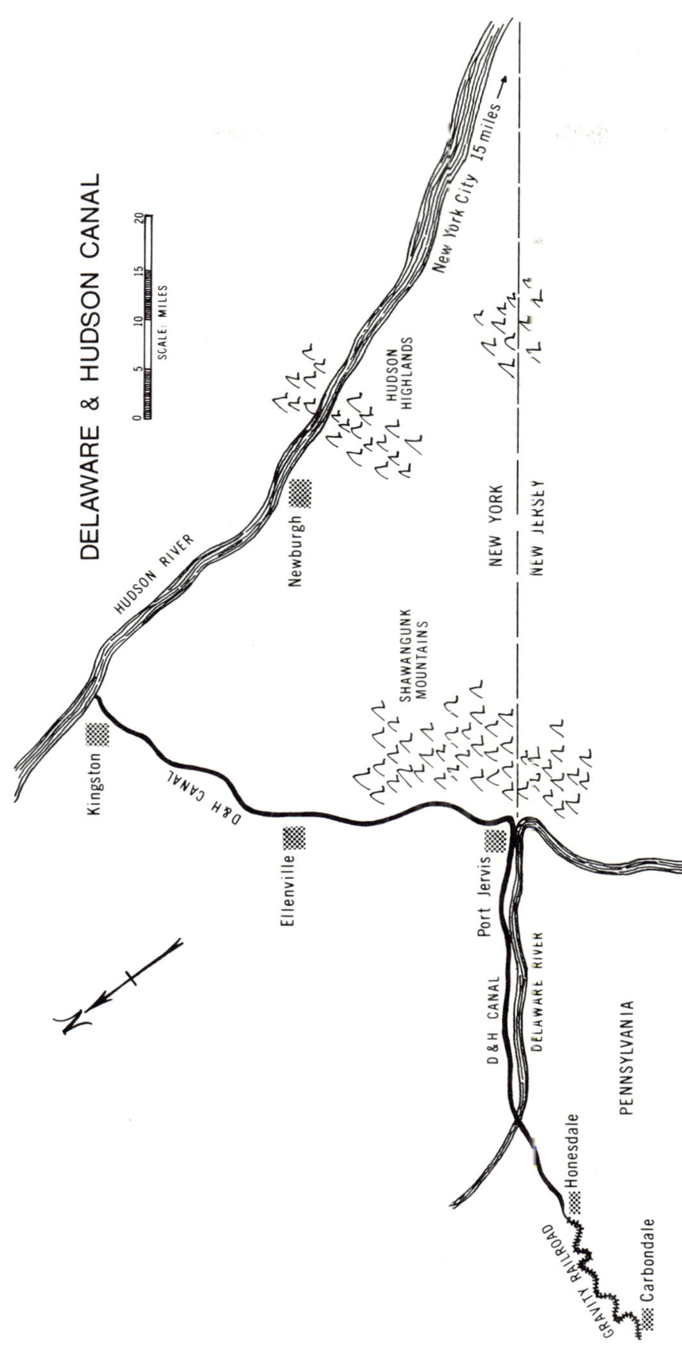

Sketch map of the Delaware and Hudson Canal.

Among the first railway passenger stations in the United States was the Crane Street Station of the Mohawk and Hudson Railway in Schenectady, New York. It was built by Jervis around 1830 and stood for a century. *Courtesy of Schenectady County Historical Society*

Drawing of Jervis' locomotive "Experiment," the first to have the famous bogie truck. Subsequently, almost all American steam locomotives were built using this principle.

The enlargement of the eastern section of the Erie Canal was planned and designed under Jervis in 1835–36. Near Fort Hunt, New York, are two features of this project. Framed in the arches of the Schoharie Aqueduct are the piers which supported the wooden, water-filled trough in which the canal boats floated; the arches carried only the towpath. Behind the remains of the Empire Lock the coping of the smaller lock built during the original construction can be seen. *Courtesy of the Historic American Engineering Record of the National Park Service*

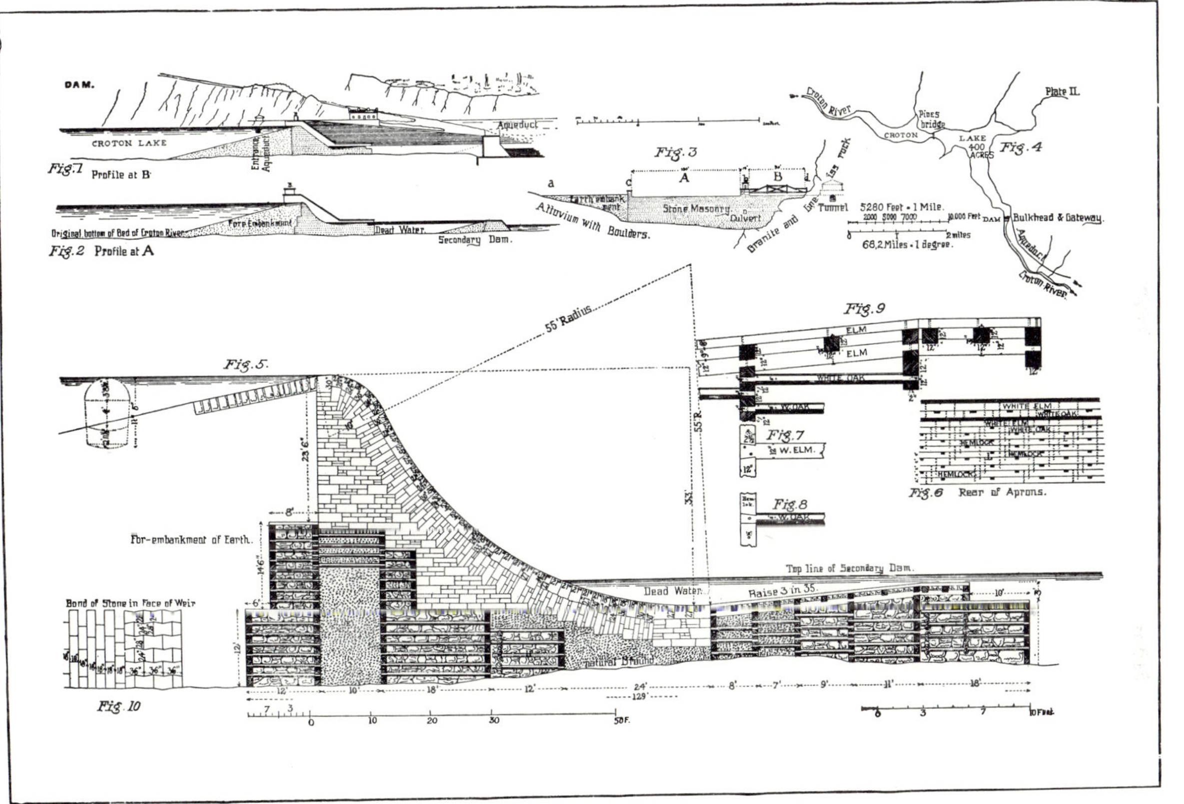

Engineering drawing of the Croton Dam, showing profiles of structure and sketch map of general areas (Fig. 4). Profiles A and B (Figs. 1 and 2) are located on front elevation (Fig. 3). The details of construction (Fig. 5) are enlarged from Fig. 2. Note the ogival shape of spillway and stilling basin (shown as "dead water"). From *The Design and Construction of Dams,* by Edward Wegman (New York: John Wiley, 1907)

View above the Croton Dam. The intake to the main conduit is located under the small building at left. It is shown in Fig. 1 of the engineering drawing. *The Smithsonian Institution*

View below the Croton Dam, as shown in Fig. 3 of the engineering drawing. The main conduit is located in the tunnel under the large building to the right of the dam. *The Smithsonian Institution*

Construction of the large iron conduit on the High Bridge. This work was done some ten years after Jervis had finished the original aqueduct. One of the original three-foot conduits can be seen under the large conduit. *The Smithsonian Institution*

Jervis' High Bridge of the Croton system over the Harlem River, during construction of the large main, viewed from the Westchester side looking northwest. The five masonry arches over the river were replaced by a single steel arch in 1937. Note, over the fourth arch from the left, the traveling crane used for erecting the large iron conduit. *The Smithsonian Institution*

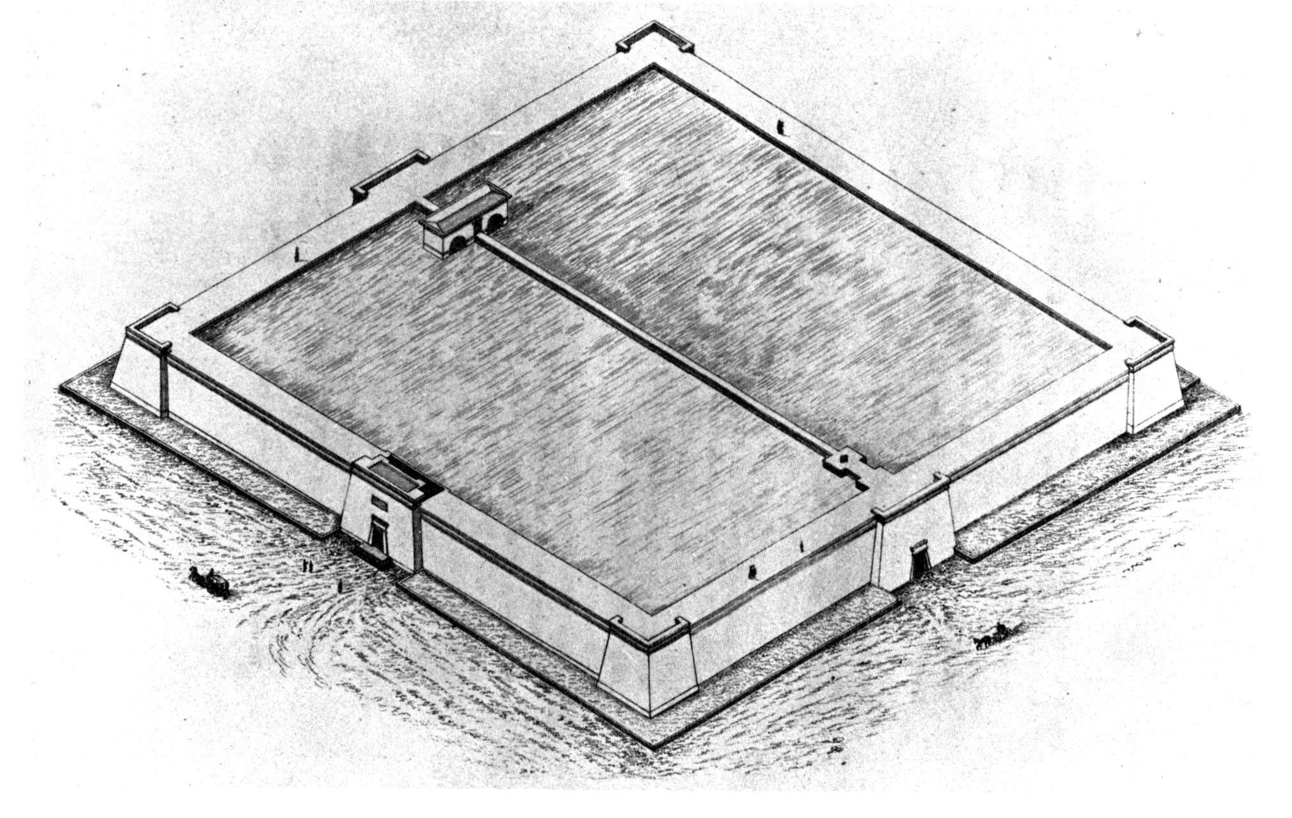

The distributing reservoir at Murray Hill was located on the site now occupied by the main building of the New York Public Library. Note the Egyptian style of architecture adopted by Jervis to relieve the dull massiveness of the high masonry walls.

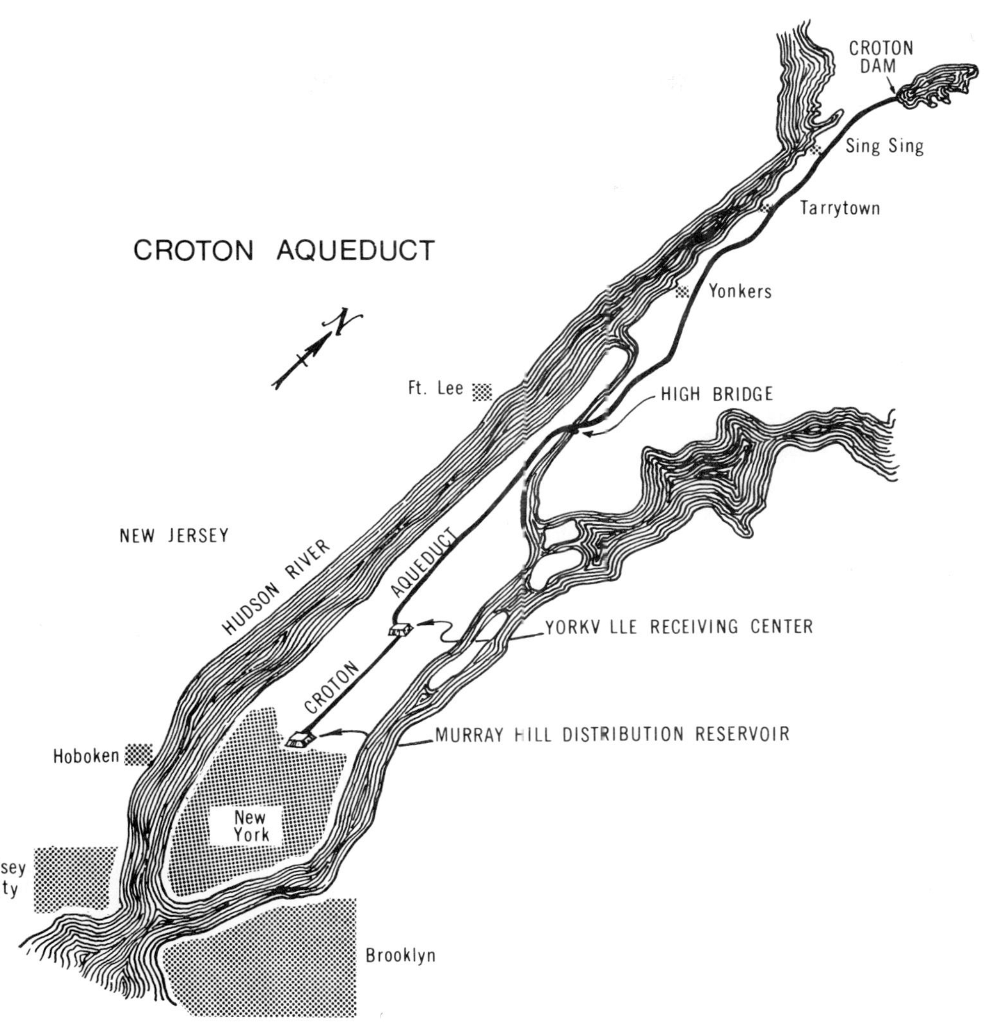

Sketch map of the Croton Aqueduct.

Early photograph of home built by Jervis in Rome, New York. Today this building is part of the public library founded under his will. Note the unique two-story brick columns. The seated couple may be Jervis and his second wife, Eliza.

these occasions, Mr. Allen and myself fully discussed all plans that suggested themselves to us to secure this object.

We both knew very well that two cars or wagons had been connected by a general frame, with transom and center pin, and used for heavy weights and long timber. That these had been used at moderate speed we well knew. We also knew that an eight-wheeled locomotive had been constructed which rested on two four-wheeled frames, so arranged as to rest on these two, and that each acted in its lateral motion independent of the other. This engine had four cylinders, two resting on one frame and two on the other, all receiving their steam from one boiler, resting on the upper or main frame. I had a plan of this engine but in some way the book is lost. Reference to it is made in Wood's last work on railways, page 136, American edition. This engine, he says, "was too complicated and failed to be useful." The question with Mr. Allen and myself was not, could two separate cars or trucks be so connected as to run on a railroad, but, could this be done in a way that would secure the requisite speed for passengers. To this we addressed ourselves, and with the utmost frankness as I believe. In our many discussions on this subject, it was clear we had different ways of reaching the object. As I have said, it was known that two four-wheeled cars had been so coupled as to carry freight, and in one instance a locomotive. But the general impression was that the arrangement would not admit of proper speed with safety. At the time I speak of, 1830–1831, I knew no engineer of much standing, except Mr. Allen, that supported or approved of my truck. Even Mr. Allen objected to it as only a half-way measure and frequently urged me to take his views. He was then preparing his plan at Albany. Mr. Allen's plan was simply putting two engines together, and coupling them by transom and pin, on which his boiler rested. He had one pair of small wheels under each engine. I was content to have a simple truck, all wheels alike, as a support, guide, and to give more steady motion. Would such an engine keep the track or rail at high speed? This was the question. The English called it "bogie" or scare-crow. The eastern railroads would not use it, and up to 1836, most of their engines were four-wheeled, English-made, or an English pattern.

You say Mr. Allen put his engine in motion one year before mine was put on. His was on about six months before mine. But this is immaterial, as all the ideas and plans were discussed together, and they were not the same in plan or result. Mr. Allen took his course, and I took mine which was a different one. His was properly a double engine with a single boiler connecting the two frames on which each engine rested separately. Mine was simply and properly a truck placed under the front end of the engine as a support, guide, and so forth. All the machinery of the engine was placed on the main frame, resting on the driving wheels at one end, on my truck at the other. This is what I claim *in idea and invention.* The fact that we took different methods is obvious from the respective practical workings.

I have shown that neither Mr. Allen nor myself is entitled to the credit of originating the truck principle, or that principle by which two cars or trucks may be connected by a common frame and made one machine, either in cars or locomotives. All we are entitled to is our respective plans for adapting this principle to passenger speed, which had not previously been done.

It is now say forty years since these two plans were put on trial. I do not know that Mr. Allen's locomotive is used on any railway. Mine may be seen on thirty-thousand miles of railway, substantially the same as put in motion in 1832. You admit there was a difference in the two, and that my application has come into general use. I can hardly ask for a more important admission. The fact is Mr. Allen's engine has never been duplicated to my knowledge. My plan is in general use.

You seem of the opinion Mr. Farlie (of England) may yet make something out of Mr. Allen's plan. Perhaps he may. If I understand Mr. Farlie's method, he makes all his wheels alike and all drivers. Now, it is clear what Mr. Farlie has to do is to devise a satisfactory connection between his boilers, resting on two cars or frames combined, and his cylinders each laid on a separate frame, having motions out of parallel. It is reported that Mr. Farlie has succeeded in this. If so, he has accomplished what I believe had not been done before.* It is clear that Mr. Farlie had based his experiments on my plan of truck, having first used it as merely a supporting element. The eight-wheeled cars have the same origin.

---

* At present (1879) we hear nothing of Mr. Farlie's double truck engine. [Still nothing in 1970.]

Not liking to trespass on your space, I have made my explanation as brief as possible.

I can bear testimony to Mr. Brown's indefatigable zeal and industry in collecting materials for a history that will be read with much interest. To look at the locomotive as registered by Mr. Brown, and then at the present condition, it is seen that a vast progress has been made, verifying the early prediction of the Westminster Review—"That the railway was an epoch in the affairs of mankind" (1833).

JOHN B. JERVIS

I have not noticed any reply to my letter above transcribed, though over eight years have elapsed since it was written.

In July, 1833, I wrote to the *Railroad Journal* [Vol. II, Part 2, p. 468] an account of the performance of the Saratoga Locomotive [the "Davy Crockett"]. No one at that time contested my claim to the invention of the truck.

As I have noticed, Mr. Allen and myself were thrown together in the summers of 1830 and 1831. Both having charge of railways, we were much interested in all devices that promised to be useful in this kind of enterprise, and especially as to locomotives. Mr. Allen had been one of my assistants on the Delaware and Hudson Canal, and our intercourse had been of the most frank and cordial character. We were providing for the railways in our charge a timber rail capped with iron. At that time it was not thought railways would be able to compete with good water conveyance on parallel lines for heavy freight; but there was no doubt of their superiority for passengers and such freight as could pay for rapid transit on lines parallel with inferior water transport.

To secure this, the railway must be operated with a speed and economy that would meet the demands of this class of traffic. We both had at that time a lively impression of the failure of the "Stourbridge Lion," and were specially intent on devising a relief for the rails by some method of increasing the number of wheels for the locomotive. The English increased the number of wheels but kept them all in one rigid frame. This we regarded as unfavorable for working on curved lines. The idea had occurred to me of supporting the front end of the engine frame on a truck of four wheels that should be able to move freely on the track and bring the engine easily around a curve by the center or transom pin. I did not doubt this would work well at low velocity; but the question was, would it be safe on curves at high speed. I suggested this to Mr. Allen, and he expressed himself as favorable to it, and as I have stated in the above letter, he was the only engineer of any considerable experience who favored my views.

When I proposed the truck, I did it as a suggestion that had been revolved in my mind for some time and held with more of anxiety than confidence. I freely accord to Mr. Allen the credit of his support in this respect. He was confident it would work safely. Our relations were extremely cordial, and we interchanged the various views we entertained in a most frank and unreserved manner. As our discussions continued, Mr. Allen's confidence in the truck led him to take his position for a double-truck engine. As this involved what appeared to me too much complexity, I did not see my way clear to follow him, preferring to adhere to my original view of providing additional support, and the security of more steady motion. I very well recollect Mr. Allen thought me too timorous, and that I would fail to obtain the full benefit of the principle.

In this situation of our intercourse, Mr. Allen proceeded to make his plan of a double-truck engine, which I think he completed his plan in the summer of 1830. If this engine had proved a success, I should not feel entitled to the credit, though I had first proposed the truck. All the responsibility rested on him, and I should have rejoiced at his success. At the time I had doubts of the successful working of the double-truck engine, more especnally as it involved the working of machinery on two frames that must be constantly varying their parallelism. It appeared to me there would be difficulty in keeping the movable steam joints in order. But Mr. Allen had confidence he would overcome this difficulty.

I was not at this time ready to make any plan, nor did I put my plan in form until the next year. In his visit in 1831, he continued to urge me to adopt his plan, which was then in process of construction. But I could not get over my doubts, and preferred my first idea, to try the truck for the front end of my engine. As I have stated the anxiety was as to its working well at high speed. The fact is, it was at that time a pretty bold experiment for either of us, and Mr. Allen was rather more bold than I was. If Mr. Farlie succeeds, as stated, in making the double-truck engine a success, it will be carrying out Chapman's [16] plan of a double-truck engine. Mr. Allen adopted his plan of engine, but of late I hear nothing of Mr. Farlie's progress in this matter.

In all my discussions with Mr. Allen I steadily adhered to my original plan of a truck under the front end of a locomotive, while Mr. Allen held to his double-truck engine. Both have been put on trial, as to the distinctive features of each, and while I have no claim to Mr. Allen's double-truck engine, I cannot concede to him my own plan or even the idea of the plan which Mr. Allen said was only going half way.

The first trial of my truck locomotive [the "Experiment"] was made on the Mohawk and Hudson Railway in 1832 by David Matthews,[17] a machinist from the West Point Foundry Association, subsequently the Master Machinist of the Utica and Schenectady Railway. He ran the engine with a view of testing the safety of the truck. After experimenting with it until he made a speed of fifty miles per hour, he very confidently pronounced the truck perfectly safe for any speed a locomotive should run.

The truck engine was not approved in England in 1850,[18] though it had been successful on American railways nearly twenty years with all varieties of speed required for a locomotive.

16. This may possibly have been William Chapman, an English inventor and railroader who, in 1812, patented a bogie truck device which was recommended but not adopted by R. Stephenson in 1828.

17. David Matthews had also supervised the building and testing of the "DeWitt Clinton" which was delivered to the M&H in June, 1831, and put in service on August 9.

18. The English engineer J. V. Gooch had actually adopted a bogie truck design in 1849, as noted in "Landmarks in Steam Locomotive Design," in *Engineering Heritage,* by W. D. Skeat (London: Institution of Mechanical Engineers, 1963).

To the above I make one exception. When traveling in England in 1850, I came across a truck locomotive on the Berwick Railway. This railway was unusually crooked for an English railway, and the superintendent said to me they had put on a few of this kind, thinking they would work easier on the curves, and as they did not run high speed they concluded to risk them; and that they found them to run very well. I mention this as showing the English apprehension of this class of locomotive. Subsequently, as I have stated, Mr. Farlie has introduced the truck locomotive, and although it had been some thirty years in successful use on American railways, the English now claim the truck as an English invention, which it certainly is not; it is American. The general impression I found in England was against the truck locomotive as unsafe, and I cannot see how they can now claim as English a plan so long in use on American railways.

I have been informed that Mr. Baldwin,[19] the Philadelphia locomotive builder, at one time claimed the truck as his invention. The benefit of the truck has become so obvious that we need not be surprised that it has so many claimants. The successful application of the truck to locomotives soon led to its adoption in the double-truck car and coach, now in general use on American railways. This was simply to duplicate my truck, which was found to work well on a locomotive. Mr. Winans took out a patent for an eight-wheeled car on this plan; but as it was held [by the United States Supreme Court] to be a mere duplication of a truck already in general use, he did not, I think, obtain much benefit from his patent.

For a long time, and now on some railways, I was acknowledged as the inventor of the truck, and complimented with a free ticket on the railway. Under the administration of Mr. Corning I enjoyed this freedom on what is now the New York Central and Hudson. But Mr. Vanderbilt does not know me, and though I daily see on his trains in the truck and other devices I have invented and put in successful operation, he refuses me

19. Mathias W. Baldwin began his career as a jeweler in 1819 and in 1831 designed and built a small demonstration engine for the Peale Museum in Philadelphia. His first major locomotive was "Old Ironsides," a four-wheeler for the Philadelphia, Germantown and Norristown Railroad, delivered in November, 1832.

such compliment, which I think my devotion to railway affairs entitles me to. The railway fare that I pay is of no great importance, but I think I am entitled to the compliment, and as I see his trains pass I instinctively feel that I am not justly dealt with. On the Hudson River portion of this railway I have a life pass for myself and wife, and can ride there in the enjoyment of a privilege I have earned by hard labor.

THE AMERICAN SWITCH RAIL

I employed Mr. Asa Whitney, an experienced mechanic, to take charge of the construction of the machinery for the inclined planes on the Mohawk and Hudson Railway and such incidental work as appertained to the other items of machinery. In this service Mr. Whitney invented and put in operation what may be called the American Switch Rail. Instead of switch points as used on English, and I think on all other railways at that time, Mr. Whitney substituted a joint or movable rail, which allowed a much more favorable angle in changing the direction of the train. In 1850 there was an attempt to go back to the English points, but after a short trial, the general practice returned to the Whitney switch. There was some advantage as to safety in the English points, as when a switch was placed wrong it did not necessarily throw the train from the track; but on the whole, the American switch was found to be superior and is in general use. When I left, Mr. Whitney was superintendent of the operating department of the railway. Subsequently he became one of the Canal Commissioners of the State of New York. After the expiration of his time of service as Canal Commissioner, Mr. Whitney entered into partnership with Mr. Baldwin of Philadelphia in his locomotive works. Mr. Whitney dissolved this connection in a year or two, and went into the manufacture of car wheels which, on a large scale, he carried on very successfully until his death.

The Mohawk and Hudson and the Schenectady and Saratoga railways were completed and I resigned the situation of Chief Engineer in April, 1833.

VI

MOHAWK AND HUDSON RAILWAY

AFTER the partial close of my engagement on the Delaware and Hudson Canal and Railway, in the spring of 1830, I made an engagement as chief engineer with the Mohawk and Hudson (Schenectady and Albany) Railway.

This railway was projected for the transportation of passengers, though it had the right to transport freight also.

Hitherto little had been done in railways for general traffic. In the year 1825, the Stockton and Darlington Railway in England was opened for general traffic in addition to coal traffic, the latter being the main object.[1] A small amount of general freight and about fifty passengers per day constituted its traffic exclusive of coal in 1827. It may be regarded as the first railway that attempted general traffic. I think this was the first railway that was operated by the mixed system of locomotive and stationary steam power to any great extent. The locomotive success on this railway was the most important practical stimulus to locomotive enterprise at that day.

In the year 1829, the Liverpool and Manchester Railway in England was completed. This was designed for general traffic of passengers and freight. It must, I think, be regarded as opening the epoch of railways, which has revolutionized the commercial and social intercourse of the civilized world. In October of that year, the celebrated [Rainhill] trial of locomotives was made on this railway. It is curious to notice the fact that the engine

1. On September 27, 1825, the official opening of this railway took place with George Stephenson's "Locomotion" pulling a string of twenty-one coal wagons and her sister engine "Experiment." About three hundred persons sat on benches fitted in the wagons.

"Rocket," constructed by George Stephenson, weighed only four and one quarter tons. All the others in this trial but one weighed less.[2] The important and I may say essential peculiarities of the "Rocket" were the steam blast and the tubular boiler. These constitute essential elements in the locomotive engines of the present day. Some important additions have been made in American locomotives, which will be noticed hereafter. The steam blast [3] I believe was introduced on a locomotive in the vicinity of Newcastle in England, I do not know by whom. The tubular boiler [4] was suggested by Mr. Booth, the Treasurer of the Liverpool and Manchester Railway. It has been supposed by some that both these improvements were devised by Mr. George Stephenson. Though it does not appear he was the inventor, his practical sagacity readily seized the principles of both and incorporated them most efficiently in his locomotives. He is therefore entitled to the credit of bringing these great principles into practical operation and thereby giving date to the railway epoch, more specifically announced in the Westminster Review in 1833.

At the time of the trial above referred to, it was not known how far the adhesion of the wheels to the rails could be relied on as a means of traction. The old methods by cogwheel [e.g., the Bleckinsop/Murray Locomotive of 1812] and other devices were abandoned as inadequate for long railways. Stationary and locomotive steam engines were regarded as the only suitable means of transport on railways. On inclined planes stationary engines were regarded indispensable and locomotives for level grades or those of light inclination.

2. The Rainhill entries were (with unloaded weight): "The Novelty" (three tons, 15 hundredweight), Messrs. Braithwaite and Erickson; "The Sans Pareil" (four tons, five hundredweight, two quarters), Mr. Hackworth; "The Rocket" (four tons, three hundredweight), Mr. R. Stephenson; "The Cycloped" (three tons), a horse-treadmill, Mr. Brandreth; and "The Perseverence" (two tons, seventeen hundredweight), Mr. Burstall. Only the first, second, and third were serious entries. It should be noted that four tons, three hundredweight are in British units and equivalent to 11,536 lbs. or almost five and three-quarter tons American.

3. The "steam blast" is a system of ejecting waste steam through the smoke-stack to increase draft.

4. Early locomotives heated water for steam using oversized kettles; later, water was circulated through tubes for more efficient heating.

On the line of the Liverpool and Manchester Railway [5] (exclusive of the inclined plane at Liverpool) there were two planes of one and one half miles each, with a rise of one in ninety-six, or fifty-five feet to the mile. It had originally been designed to work these planes with stationary engines, but Mr. Stephenson had so great doubts on the expediency of this that no machinery had been provided up to the time of the trial above mentioned. It was with a view to settle this question the trial of locomotives was instituted, as well as to call out the mechanical skill of the country in producing them. Previous to this trial the company submitted the question of stationary or steampower for the inclined planes to engineers of eminence, who gave much time to the investigation, treating the subject as applicable to the light grades as well as the inclined planes, and came to the conclusion that stationary power would be the best. Mr. Stephenson and Mr. [Joseph] Locke were in favor of locomotive power.

In the great competing trial above referred to, one of the conditions was that the weight of the locomotive should not exceed six tons and that it should rest on six wheels. The limit of weight for a four-wheeled engine was four and one-half tons. The "Rocket" as I have before stated weighed four and one quarter; and the other principal competitor, the "Novelty," weighed three tons, seventeen hundred weight [8,624 lbs.], carrying her own fuel and water without a tender, as usually attached for that purpose. It is as historical matter I mention these things, for it is not easy for men of the present day to realize the great progress that fifty years has made in the march of the locomotive or to consider those inventions of half a century ago that laid the foundation of this progress. It was then a grave question on which the engineering profession was seriously divided, whether a locomotive could profitably be used on a grade of fifty-five feet to the mile. Now such grades or even one hundred feet to the mile would raise no such question.

The Honorable Stephen Van Rensselaer was the first president of the Mohawk and Hudson Railway. John Jacob Astor,

5. The Liverpool and Manchester Railway was opened formally on September 15, 1830, the lead train piloted by R. Stephenson.

Lynde Catlin, J. Renwick, and C. C. Cambreling were prominent directors.[6] The latter was made the general agent or commissioner, and John I. De Graff of Schenectady his assistant.

The citizens of Albany were not generally favorable to this railway at the time of its commencement. To this I must say there were some eminent exceptions. This unfavorable sentiment in Albany was opposed to any location in the streets of the city, and this led to the location of the line to the south side of the city. The Albany and Schenectady Turnpike Company, the stock of which was principally held in Albany, was the nucleus of this opposition. They commenced to make a stone track on this turnpike with a view to sustain competition and impair the value of the railway. It was boldly asserted the sand from the cuts through the sand ridges would be so driven by winds into the railway as to destroy its usefulness. The railway company, however, did not lose courage, and they steadily persevered in the enterprise.

Though it was intended to use locomotives between the inclined planes, a horse path was formed in the first instance and used more or less until the locomotives were fully prepared.

The railway was opened for the transportation of passengers, partly by horse power and partly by locomotives, in August, 1831. After this, the company decided to lay down a second track.

After the railway was opened for transportation the citizens of Albany did not like to go to the station at the south end of the city to meet the train and petitioned the company to make a branch line from the junction of the railway and western turnpike along the turnpike to State Street and down this street to a point in front of the Capitol, to be run by horse power. The

6. John Jacob Astor was a German immigrant who made a fortune in the fur trade of the Northwest. At the time of his death in 1848 he was probably the richest man in the United States. Lynde Catlin was an original "manager" of the Delaware and Hudson Company and served from 1825 to 1826. Churchill C. Cambreling was a longtime (1821–39) Democratic congressman from New York and was later minister to Russia (1840–41). He was also first president of the Saratoga and Schenectady Railroad, breaking ground for it on August 20, 1831.

railway company granted this petition and constructed the branch. As soon as the branch was completed, the passengers were put on it and the old station at the south end of the city was used only for freight.

As the branch line was allowed to use horses only, it was found to be an inconvenience in the operation of the railway, and after a few years, when horse power was suspended on all other parts of the railway, the company abandoned the use of the branch and confined their operations to the main line, which was unsatisfactory to Albany.

For a general description of the main features of the route of this railway, I make an extract from my *Memoir:*

This line extended from Albany on the Hudson to Schenectady on the Mohawk. Its main object was the transportation of passengers. At that time passenger boats on the Erie Canal brought passengers from Buffalo to Schenectady and together with stage coaches transported a large number.

The intermediate country was a table land of fair character for a good line and easy grades, but that was reached by a sudden rise of near two hundred feet from the Hudson and over one hundred feet from the Mohawk. These rises were at that time regarded as requiring inclined planes worked by stationary engines. Consequently this railway was constructed with an inclined plane at each end. The adhesive power of the engine wheels to the rail was the means of transmitting power from the locomotive engines and was then regarded much below what it has subsequently been found to be. After working with the inclined planes a few years, this inconvenience for a miscellaneous traffic was demonstrated and a commission was appointed to ascertain the practicability of dispensing with them. I was a member of the commission and examined the engineering view of the first question. We found a line rising from Albany at a grade of eighty feet per mile and from Schenectady at forty feet per mile which the commissioners recommended, and the line was accordingly changed in this respect and has subsequently been worked by locomotive steam power.

This railway was constructed with a double track. It had the ordinary features of railways at that time constructed in this country, namely, a wooden rail capped with a plate of iron, which in the excavations was laid on stone blocks and on embankments with timber cross ties. While in use the inclined planes worked very smoothly, and I believe no serious accident occurred.

A locomotive engine was made for this line by the West Point Foundry Association. It weighed between four and five tons, had four wrought-iron wheels, and carried from seventy-five to one-hundred and twenty-five passengers in a train, at a speed of twenty-five miles per hour. A second locomotive was obtained from the works of Robert Stephenson, England, which weighed about seven tons, and was much more powerful than the other. The first was named "DeWitt Clinton" and the second "John Bull." [7] The action of the "John Bull" was observed to be rather severe on this kind of rail. It being placed on four wheels, the overhanging caused a sharp and disagreeable motion of the engine. This circumstance, with others which I have noticed in previous observations, induced me to continue my researches for a remedy for the weight and secure a more steady motion for the engine. The idea of more wheels was natural for the first object, but to spread the base was supposed unfavorable for curves; and I was finally led to the plan of a four-wheel truck under the forward part of the engine as a support for that end and as a guide on curves. By this I had four instead of two wheels, and could place the truck so far from the drivers as to reduce materially the vibratory motion. A new engine [the "Experiment"] was built by the West Point Foundry Association on this plan, and the truck was found to work satisfactorily. I then made a new plan for an engine for the Schenectady and Saratoga Railway (of which I was the chief engineer), and sent it to Mr. Stephenson, who built the engine. This engine weighed about six tons; it moved with perfect beauty, and I may be pardoned I think in saying that my first ride on it was a great delight. The smooth action of its parts and the steadiness of its motion gave me great

7. According to the *History of Transportation in the United States before 1860*, ed. by B. H. Meyer (Washington, D.C.: Carnegie Institution, 1955), p. 357, the first two locomotives were the "DeWitt Clinton" and the "Robert Fulton." Presumably "John Bull" was the original, that is, Stephenson's, name for the "Robert Fulton." The Camden and Amboy had purchased a locomotive also called the "John Bull" from Stephenson in 1831. In 1833, Isaac Dripps, the railroad's master mechanic, fitted this locomotive with a pair of pilot wheels which then supported the first "cowcatcher."

satisfaction. A year or two later, the railway was extended westward, and from time to time continued, until it reached the Pacific Ocean at San Francisco. On the entire route, the truck arrangement was considered a complete success, and is now the general plan of American locomotives. The truck itself was not a new idea with me; it had in principle been used, but was generally supposed to be impracticable for high speed.

VII

CHENANGO CANAL AND
ERIE ENLARGEMENT

AFTER completing the service above mentioned, I made an engagement with the Canal Commissioners of the State of New York as Chief Engineer of the Chenango Canal [April, 1833].
In my *Memoir*, I give a brief description of this work:

> The Chenango Canal is about ninety-eight miles long and has about one hundred locks. The cost was about two million dollars; it extended from Utica on the Mohawk River through the Oriskany and Chenango valleys to Binghamton on the Susquehanna. The route was moderately favorable. The most distinctive feature is the resort to artificial reservoirs to supply its summit with water. At the day this canal was commenced, it was believed no such dependence in this country was relied on to supply canals in Europe, and it had been held that one-third of the rainfall could be depended on over the loss of absorption [by the soil] and evaporation. As this ratio would vary in different situations, I decided to establish a rain gauge [1] and a sluice for measuring the water that flowed from the valley. Mr. William J. McAlpine was the resident engineer on the summit section and had charge of the rain and sluice gauges. The matter was diligently attended to, and the result from these gauges was to establish forty percent as the proportion of rainfall which on an average could be secured in a reservoir.

This canal had not much promise of commercial success. Several surveys had been made, all tending to show it could be made very cheaply. The project met with strong opposition

1. This coordinated measurement of rainfall and runoff is a hallmark in the history of American hydrology. Even D. H. Mahan's classical text, *An Elementary Course of Civil Engineering,* mentions only stream gauging.

which for several years was too powerful for its friends; but the resolute persistence of the latter finally prevailed, and the Canal Commissioners were instructed to proceed with its construction. The opponents of the canal laid considerable stress on the difficulty of obtaining sufficient water for its summit.[2] The criticism of these opponents to the canal did not cease when its construction was authorized. So far as they charged as to failure of sufficient traffic they had a strong basis. But the criticism on the supply of water for the summit was not well taken. By way of ridicule it was said, "Jervis is going to supply the Chenango Canal through brass cocks." The cocks however proved quite adequate to the duty that was expected from them. Plans of the reservoirs and their appurtenances I have deposited in the library of the American Society of Civil Engineers. They were a complete success in supplying the summit with water.

I applied the same calculations as to drainage in my report on the Cochituate Aqueduct for Boston, where they have been fully verified.

ENLARGEMENT OF THE ERIE CANAL

From my *Memoir*, the following is an extract:

The Canal Board had decided to enlarge the section of canal to five feet depth and fifty feet width. This decision was not fully approved as being adequate to the wants of the future traffic, and the matter was earnestly discussed by the public, which caused indecision and suspended operations. My connection with Commissioner Bouck on the Chenango Canal led us to much conversation on this matter, and at his request, I investigated the question of the ratio the section of boat should bear to the section of canal, deducing the comparative economy of transportation. Without dwelling on preliminary facts, I may state these discus-

2. Perhaps the most famous American failure ascribed to this problem was the Santee-Cooper Canal in South Carolina. Had it not failed, it would have been the first successful inter-basin canal in the United States. An inter-basin canal connects two separate river watersheds, and thus must cross a topographical divide. This is in contrast to a bypass canal, which connects two portions of the same river, circumventing falls or rapids.

sions led the Canal Board to order a careful survey of the work, and estimates to be prepared on the basis of enlargement to a depth of six feet and also seven feet of water, the widths being respectively sixty and seventy feet.

In accordance with the above, engineers were appointed to make this examination on the respective sections as directed by the Commissioners and to the writer the eastern section was assigned. The surveys and estimates were made in 1835. The surveys of this [eastern] section were made by W. J. McAlpine, who at the time was one of the resident engineers on the Chenango Canal. Though I was in charge as chief engineer of the latter canal, I gave a large share of personal attention to the question of enlargement.

As I have stated there was much discussion on the question of enlargement, and before the survey was ordered as above, the question of a ship canal from Oswego to Albany was presented to the legislature. This was referred to the canal commissioners, and by them referred to three engineers in their employment, viz.: Holmes Hutchinson, Frederick C. Mills, and the writer. The project of the Erie Railroad was then before the legislature, and its friends urged that railways would supersede canals, and consequently no enlargement of the Erie Canal was necessary. The legislature referred these matters to the Canal Commissioners, and they again to the engineers above named. On the engineers devolved the labor of procuring data and writing reports on the two subjects. That on the comparative value of railways and canals was presented 14th March 1835—Assembly Document 296. That on the ship canal on the 23rd March 1835—Assembly Document 334. The views presented in these reports were approved by the Canal Commissioners as the basis of their report to the legislature. They had incidentally a bearing on the question of the enlargement of the Erie Canal. At present those documents are only of historic value.

The railway had not reached the development of the present day and was not then supposed capable of reaching it, though conceded to have great advantages for general commerce. The first real start of the railway for general traffic was when it was discovered the power of a locomotive steam engine could be effective through the adhesion of the wheel to the rail. Engineer [Nicholas] Wood made experiments on this about 1822, and claimed that one twenty-fifth of the insistent weight of driving wheels on the rail could be relied on as a tractile power. The experiments on the Liverpool and Manchester railway proved this ratio was too low. The ratio has grown until it is now conceded to be one-fifth or five times the ratio Mr. Wood found as a safe one. This is the secret of the great advance in locomotive power. No doubt there has been improvement in machinery, but this principle of adhesion is the foundation of railway economy.

The reports of the engineers, Messrs. Roberts, Hutchinson, Mills, and Jervis, of the surveys and estimates for the enlargement of the Erie Canal were submitted to the commissioners near the close of 1835, and may be found in Assembly Document No. 99, 26th January 1836.

Mr. Mills and myself did not make any recommendation in our reports as to the size we regarded most suitable, not considering we were called on to do so. But when we found the others had given their opinion, we made a volunteer report to the Commissioners on this point. This report will be found under letter E. of the document last mentioned. We decidedly recommended the depth of water eight feet, and the surface width eighty feet. This had been my opinion for some time previous.

The subject being now fully before the Commissioners and the public, and having been very fully discussed, the Canal Board decided to adopt seven feet depth and seventy feet surface. At the same time they decided to make the lock walls two feet high above the upper level of canal. This allows the water to rise to eight feet, and only requires the banks to be raised to a corresponding height. The latter may be done at no serious expense, and thus secure a depth of eight feet, which would be decided improvement to the navigation. At the same time the canal board decided to construct double locks on the eastern division.

I was appointed Chief Engineer of the eastern division. William J. McAlpine was my principal assistant. Plans, specifications, and forms of contract were prepared and submitted to the canal commissioners, also a revision of the line of canal at several places. Early in 1836, the commissioners decided to proceed with the construction of the work of enlargement.

From my *Memoir* I make the following extract:

It had appeared to me that errors had been made in the original work which it was important to correct so far as practicable in the enlargement. At a point about eight miles above Albany commenced the location of what was termed the "nine locks." In a distance of about two miles there were I think seventeen locks. At one cluster the nine and at another the "four locks" had very contracted pond reaches, which was found to be very unfavorable to the large traffic that then existed. I directed Mr. McAlpine to commence a survey at a point one mile below the "nine locks," to run the line above the "four locks," and ascertain if it was practicable to find ground on which these locks could be located so as to give a pretty uniform space for the pond reaches between them. The ground was pretty rough and rocky, but a satisfactory location was found, allowing the locks to be very evenly distributed on a new line about three and one half miles in length. I therefore recommended the commissioners to abandon the old for the new line; this was done and the new line constructed, which has proved a very satisfactory improvement.

A short distance beyond the point above referred to the canal crossed the Mohawk River to the north side, was continued on that side twelve miles, and then crossed back to the south side of the river. This involved two aqueducts over the river, averaging each nearly one thousand feet in length. I had a line continued on the south side of the river, with a view to judge of the propriety of dispensing with the aqueducts. The surveys and estimates induced me to recommend the abandonment of the canal on the north side, and of course the two aqueducts, and to adopt the new line on the south side of the river. On this question the canal board was equally (6 to 6) divided, and of course the recommendation was not adopted.

The next point of importance in changing the original plan was on a section from Schoharie Creek to a point four miles above. On the west bank of Schoharie Creek the canal was locked into it, and by means of a dam, a pool was made as the channel of the canal across the creek. At a point four miles above was another lock, and about midway on this section a pretty large creek, though much smaller than the Schoharie, crossed the canal by means of a dam. These were turbulent streams and often the source of much vexatious trouble to the navigation. I proposed to take up the two left locks by which the level of the canal was raised sufficiently to cross over both the streams by aqueducts. This change the canal board sanctioned, and it has been a great benefit to the navigation.

The plan of stone arches for the towing path on aqueducts, where the height required a timber trunk for the boat channel, was proposed by me at the same time and has been adopted in the subsequent works on the canal.

Early in the year 1836, the canal board decided to proceed with the improvement as above stated and in the eastern division to make double locks.

The several engineers, Messrs, Hutchinson, Roberts, Mills, and myself, were instructed to prepare plans and specifications, and forms of contract for the works required. As this section or division was allotted to me, I had more especially devoted myself to this duty. The plan I submitted for double locks was essentially adopted, but not fully the specification. I had provided the face walls should be of large stone, well dressed to a close joint. So far they were accepted. The rear or backing stones to be large, well-shaped quarry stone, well bedded in hydraulic mortar.[3] Instead of this, it was proposed to make the backing of dimension stone more roughly cut, involving the expense of getting out the dimension or backing stone by keys and wedges. This latter the commissioners adopted. I had no desire to do anything short of what would make the work of the most stable and durable character. This work in the backing I regarded as unnecessarily expensive. But as the old locks of the canal had proved less stable than was desired, and as the Erie Canal at the time was regarded as requiring the best possible work, superfluous expense was considered a merit. The characteristic of

3. A mortar which will harden under water, consequently, is less likely than ordinary mortars to leach out from intermittent exposure to water.

the superlative was carried to a large extent in other masonry, largely increasing the cost with no benefit to the works. The face of a lock should be of large stone well bonded and cut to a close joint—the backing as a proper support to the face may be equally substantial if made of large quarry stone well bonded with the face and laid in its bed with hydraulic mortar. I still think the commissioners were in error.

From the *Memoir* before noticed:

No one appreciated more than I did the importance of substantial and durable works; but unfortunately at this time, there was so high an estimate of the value of the canal that the ideas of men were extravagant and advocated work of an expensive character that was in no way more substantial or useful. This induced a more expensive policy that increased the cost of the work much beyond what was necessary. I had recommended that the chamber of the lock for a canal seven feet deep and seventy feet wide, should be sixteen feet wide and one-hundred and fifteen feet between the gates. The canal board on the petition of navigators increased the width to eighteen feet, which I regarded a decided error. A navigation with few boats may be conducted on a comparatively narrow channel; but for a large traffic, there is a proper relation between the area of boat and area of channel that secures the best economy in transportation. The boatman will make the boat as wide as he can pass the locks, with simply room to pass each other on the open canal. They analyze nothing, but suppose the acme of economy is in the largest possible load they can carry, very much on the theory of most railway superintendents who consider the largest possible team as securing the best economy in transportation, with no more attempt at scientific analysis than that of the boatmen. It is now the opinion of the most intelligent navigators of the canal that the locks are too wide for the best economy of transportation.

After the enlargement of the Erie Canal was completed, the traffic became very large and it was generally supposed it could have no serious rival. The event has proved how little we then knew of the growth and economy of railways as of other efforts of skill and enterprise to supersede what we had supposed proof against serious rivals. In chartering the companies to construct railways along the Erie Canal, the legislature provided against their interference with the canal traffic by imposing canal tolls on all freight carried by the railways. To the railways were left only the unrestricted transport of passengers. This restriction on freight was in a short time made applicable only to that carried during the season of navigation. The public, feeling very secure in the power of the canal to control the great mass of the traffic, the railway was emboldened to ask release from the restriction at all seasons. It was urged by the commercial interest that such relief would benefit commercial intercourse in the facility it would give for more rapid transit of goods that could afford to pay an extra charge for the benefit of dispatch. This was urged on the basis that heavy freights requiring economy would continue to follow the canal and that its tonnage would not be materially lessened; that the benefit to commerce by giving greater facilities to light merchandise would increase the canal traffic in heavy goods, and so the state would not be materially a loser in tolls. On these views the state released the railway and gave them unrestricted competition with the canal in the transportation of all freight.

The sequel has shown that the power of the locomotive engine on the railway was not then understood. This has been increasing, and as it progressed it arrived at a stage that brought it in competition with canal transport on heavy as well as light freight. The first influence was to compel a reduction of canal tolls, and this was continued from year to year until the large revenue of the state was so reduced as to be hardly sufficient to pay the expense of maintenance. I think in one year the expense exceeded the tolls. This condition of affairs led to the discussion of the measure of removing all tolls from the canal, or making it free. It therefore appears the canal which at one time maintained a high commercial prosperity for the state, furnishing a large surplus revenue, was reduced to a struggling competitor with railways that had grown up under the liberality of the state, undermining their own revenues.

The question is a serious one to the state, independent of the revenue from tolls on canal freights, for if the canal has no superiority over railway transport, the city of New York will have no superiority over other Atlantic cities in controlling the trade of the western states. To the State of New York this is an important question. The railway has grown, as I have stated, and will not go back, and the question is to consider whether the canal can in any way be so improved as to meet successfully this competition and others that are impending.

In 1877 an article appeared in Scribner's *Magazine* setting forth the Erie Canal as in a state of decadence and anticipating an early day when it would have no other interest than as a ruin. Though far advanced in years, I did not read that article without feeling a prompting to examine if there was not some way in which the calamity might be averted. My reflections on the subject were published in an article in the *International Review* of May, 1878.

My article considered three points, namely, the substitution of steam for animal power for towing boats, bringing an additional depth of one foot to the water in the canal, and a more businesslike process in the management of repairs.

In regard to the use of steam in towing boats, I was satisfied this was necessary to fully obtain the best results for the canal. It is beyond doubt the great advantages in the use of steam on the railway that gives it its peculiar superiority over the canals. In this it competes successfully with animal power on canals. At a suitable velocity, freight will move on a canal with less power than on a railway. For any freight that requires a high velocity, the movement is less economical on the canal. The bulk of freight, within reasonable limits, seeks cheapness more than speed, and in this the canal has its superiority. The objection to steam on canals is that it requires system in its operation. The boats must be moved in fleets, requiring them to be brought together to form the fleet. In the case of animal power, each boat is moved independently of all other boats. There is also the objection to creating a corporate company who shall have the whole business of towing boats in their hands. To carry this plan of steam towing into operation, the towing party should be subject to regulations by the canal authorities who need have no difficulty in a well arranged system of business. The system requires a rail track on each bank of the canal. The expense of constructing and maintaining these tracks could be largely reimbursed by the local and winter traffic they could do greatly to the benefit of the local trade along the route and for through business during suspended navigation by frost.

For the purposes of traction, there is no known process of using steam as economically as that by a locomotive engine on a railway; and this may be used to tow canal boats as well as railway cars.

In regard to the second, the addition of one foot to the depth of water in the canal, I am glad to be informed that the canal authorities are now seriously considering the question. Also in connection with this, they are engaged in plans to improve the handling of the locks in the expectation of reducing the time of passing a boat to nearly or quite half what it now required. These two measures will materially improve the economy of transportation, whatever be the power used for towing.

With regard to the third question, namely—the more economical management of repairs—it is believed important progress has been made during the past year. The only question is can the state conduct the repairs of the canal on a business basis. It is hoped they may do this and save the state from the waste of political management which has cost millions.

With these three points provided for, boat transportation will be so cheap that no rival will be able to destroy the traffic, and the Erie Canal may be expected to exercise a controlling power in the trade of this and the western states. I expect all this will be accomplished, though adverse interest and old habits will probably delay the consummation.

VIII

CROTON WATER SUPPLY SYSTEM

The Croton Aqueduct

In September, 1836, a committee of the Croton Aqueduct Commissioners called on me with a proposition that I should accept the position of Chief Engineer of that work, and in October following I accepted the position. The Canal Commissioner (the Honorable William C. Bouck) with whom I was especially associated was reluctant to have me leave the canal but said he would not object to my accepting the charge of the aqueduct if I thought it my interest to do so. I have been charged with intrigueing for the engineership of this work, but that is an entire mistake, as I had not mentioned the subject to any one, and consequently was quite surprised at receiving the proposition. All the intercourse I had on the subject was an answer to a letter of the chairman of the Croton Aqueduct Board, requesting me to send him any copies I had of the forms of contract and specifications of the state canals, in which there was not the least allusion to my having anything to do with the work. At first I declined the offer, not feeling willing to interfere with Major Douglass,[1] who, I knew, had a high estimate of the professional importance of the work. But the committee placed the matter on the distinct ground that they had decided a change was necessary—that I was preferred, and if I did not accept they should seek someone else, and they wished me to consider Major Douglass as out of the question. Under such circumstances, I saw no impropriety in

1. David Bates Douglass was a graduate of Yale. For his heroic action as an army offcer in the War of 1812, he was given a regular commission and a professorship at West Point from 1816 to 1831. He left to become chief engineer of the Morris Canal (New Jersey) and professor at the University of the City of New York. In March, 1833, he was appointed (along with Canvass White) engineer for the Croton Aqueduct project. In October, 1836, his disagreements with the board of water commissioners led to his resignation.

accepting a position that appeared professionally desirable and offered without the least effort or knowledge on my part. Accordingly I resigned my position on the Erie Canal enlargement, and Mr. McAlpine, my principal assistant, was appointed my successor.

Surveys had been commenced in 1833, with a view to examine the feasibility of introducing the waters of the Croton River into the city of New York. Major D. B. Douglass was occupied during the seasons of 1833 and 1834, and made estimates and reports of his surveys. Mr. John Martineau was engaged on the same work in 1834. They made independent surveys and reports to the Board of Water Commissioners, and substantially agreed in recommending the Croton as the best source for supplying New York with water. In neither case was there much presented as to definite plans for the works by which the object was to be secured; these were left in general form for future elaboration. In the spring of 1835 the city of New York accepted the plan and authorized the Board of Water Commissioners to proceed with the construction of the work. Messrs. Stephen Allen, Saul Alley, Charles Dusenbury, Benjamin M. Brown,* and William W. Fox constituted the board. They appointed Major Douglass chief engineer, and instructed him to proceed to locate the line and prepare plans and specifications for the work. He made the location of the line of aqueduct from the Croton River to the north bank of the Harlem River, thirty-three miles; and determined the grade of the aqueduct at about thirteen and one-quarter inches to the mile. It was in the main, well located.[2]

After the determination of the height at the Croton and the grade suitable for the flow in the aqueduct there was not much to choose as to location between the Croton and the town of Yonkers. The country was side lying and but one general line could be adopted. The skill in location was in the choice of line across spurs of ground (and intervening valleys) jutting out from the main hill. This latter feature was frequently met, involving care as to the question of cutting through the spurs and the filling of intervening valleys.

* Soon after I was appointed chief engineer, Mr. Brown resigned and Thomas Woodruff succeeded him.

2. The preceding extract was selected by Jervis from his *Memoir.*

From Yonkers to the Harlem River, the country was more broken, and admitted of a choice of general location. The one adopted by Major Douglass appeared to be the best. After passing carefully over the line, it appeared to me as embracing the best advantages the form of country admitted. I thought the cuttings in some cases rather deep, but as to that, if it was an error, it was on the safe side, and therefore I did not propose material variations.

As before stated I entered this service in October, 1836. I found the water commissioners very urgent to prepare the work for contract. After my examination of the line as I have stated above, I reported on the location to Harlem River, with general approval. This brought us into the month of November. With thirty-three miles of line located, the commissioners could not see why they should not advertise for contracts. I was not surprised at this, for they could not be expected to see the great labor in devising plans and settling details of structures required, with specifications as to kind and character of work and forms of contract. Considering that two seasons had nearly passed since they were authorized to go forward with the aqueduct, I could but see the natural urgency they felt to go forward. But I well knew there was much to be done before contracts could be made, and as winter was at hand, when this kind of work could not generally be done, I prevailed on the commissioners to defer contracts until the next spring, allowing me the winter to make the preparation I regarded indispensable for such a work.

The enterprise of the Croton Aqueduct was an improvement for which there was no specific experience in this country or hardly any in modern times. It was hydraulic, and in this respect resembled canals; but it had peculiarities that had no parallel in canals. In short it presented at that time many features that had no specific guide from experience in this country. I commenced the work with confidence that I could bring it to a successful result; at the same time I felt the work involved very grave and difficult matters, and entered on the enterprise deeply impressed with the responsibility of the undertaking. I certainly felt the unfavorable circumstance I have alluded to in the impatience of the commissioners to put the work under contract, and was ready to accept any preparations of my predecessor that appeared satisfactory to my judgment; in this spirit I accepted some, more particularly in regard to location of the line of aqueduct, that subsequent experience and reflection has not wholly confirmed.

A natural difficulty in such works is the incredulity of the public mind on a work regarded as novel in its character and not supposed to be within the experience of the time when projected. Though the city had duly authorized the aqueduct, there were many, and some of them learned men, who doubted the practicability of its successful accomplishment. Designed to supply the daily wants of a large city with water, any material failure would be disastrous in the extreme. In a canal or a railway, any failure would only suspend traffic until repairs could be made; no comparison to the suspension of the supply of water to a large city. Notwithstanding the responsibility I felt, I had strong confidence the work could be successfully done, and set to work diligently to provide the requisite methods to secure success.

I did not hesitate to avail myself of any hint of information that I could obtain from any source that promised to be useful for the work. I think I was in all cases ready to abandon my own views in reference to any feature when the suggestions of others on a careful consideration appeared to be better. Originality was regarded as subservient to success.

I found in the office an atlas of the land maps of the line between the Croton and Harlem Rivers, prepared under the direction of Major Douglass. This had been satisfactorily done, and a few land claims had been settled. The next month after my engagement, the balance of land claims between the Croton and Sing Sing were mostly adjusted.

In the winter of 1836–37, I was occupied in preparing plans, specifications, and forms of contract, mostly for the general work.

Major Douglass had prepared a plan in cross section for the conduit. I do not know that he prepared specifications. I did not find any in the office. A few weeks after Major Douglass had retired, he sent into the commissioners a specification for a tunnel about two miles above Sing Sing. These were submitted to me. I asked the commissioners if they considered them as decided.

They then referred them to me to make such use of them as I could. As they did not conform to my views, they were not incorporated in those I afterwards prepared.

The form of the conduit Major Douglass prepared I considered in its general features good, and accepted it with such modifications as appeared to me proper.

When I came to discuss my specifications with the board of commissioners, they agreed with them, with the exception of the use of hydraulic cement in all masonry laid in mortar. Major Douglass had proposed the use of lime in the backing and hydraluic cement in the face of the masonry. One of the commissioners very earnestly insisted on the use of quick lime in the backing as a great economy over hydraulic cement. He was an experienced builder and had great influence on this matter with the board. This matter was earnestly discussed at several meetings. After exhausting what I had to say, and seeing no prospect of the board agreeing to my views, I said to them that I could not consent to the use of quick lime in any part of the masonry. It was no doubt a cheaper material but did not appear to me as affording the best security for the work, and if the board insisted on its use, *they* must assume the responsibility of the measure. This closed the discussion, and the board immediately adopted the specifications in full. Some months after, when the work was in progress, the commissioner referred to expressed to me his satisfaction that my views in regard to hydraulic mortar were adopted.

Late in February, 1837, the work of preparation was so advanced the commissioners advertised for proposals for about eight miles of the aqueduct, from Croton Dam to Sing Sing. Proposals were received to the 26th of April for this portion of the first division, and Edmund French [3] was appointed Resident Engineer. The contracts were divided into sections of about four-tenths of a mile each.

The prices bid for the work were regarded as rather high, though there was a large competition for contracts. This was not

3. Edmund French graduated from West Point in 1828, but resigned from the army in 1836 to work on the Croton project. He later did engineering work in Washington, D.C.

strange, as the work was new at the time, and the contractors felt an uncertainty as to its cost. This was made more impressive by the requirement of adequate security, strictly insisted on by the commissioners. These circumstances, while they enhanced the cost, secured a very able class of contractors. It left no excuse necessary for a rigid enforcement of the plans and specifications. At subsequent lettings of contracts, as men became familiar with the work, the rate steadily declined.

In order to secure a thorough supervision of the work—in addition to the Resident Engineer and his assistants—there were placed inspectors of masonry on each two miles of line. These parties had authority for taking down any work for examination where they had suspicion it was not properly done. They inspected all materials, especially hydraulic cement. They were required to make tests of every cargo of the latter. A committee of the commissioners with the chief engineer made fortnightly visits in the season of operations over the works during its whole progress. These latter had an important influence on all others, as no one liked to have his work wanting. It was deemed important to guard against a very common fault in hydraulic masonry, namely, the want of full bedding of the stone and brick in mortar.

The legal process that was authorized by the legislature to obtain title to the land for the aqueduct was attended with considerable delay, and this deferred the time of putting the work under contract. It was fortunate that the commissioners obtained as a board of appraisers Messrs. William Jay, William Nelson, and Abraham Miller. They were as I thought rather liberal, but this did not prevent appeals and delays by those not satisfied, and so the work was delayed until these matters of title could be adjusted.

In September, 1837, a second portion of the aqueduct was put under contract. This completed the first division and nearly the second division, total about twelve miles, or with the former, about twenty miles. On this second division Henry T. Anthony was appointed resident engineer. Mr. French and Mr. Anthony were a part of Major Douglass' department.

The third and fourth divisions extended from the south end of the second division to the distributing reservoir. These were combined for the time and placed under Peter Hastie as resident engineer. Mr. Hastie's duties for the season of 1837 were mostly confined to surveys for the location of the aqueduct between Harlem River and the terminus at the distributing reservoir, no part of the third and fourth divisions being then under contract.

After passing about one mile from the south end of the location for Harlem Bridge, the high ground fell away and the general form of the country was irregular, rendering necessary very extensive surveys in order to find the most favorable ground. The location was also embrassed [constrained] by the necessity of conforming the line as much as possible to the line of the streets. During the season a location was established and approved by the commissioners. This location involved the method of crossing the Harlem River, Manhattan and Glendenning valleys, and also the receiving and distributing reservoirs. In all these cases, important plans and specifications had to be provided. They were the most important structures on the aqueduct, and required considerable time to elaborate all their requirements.

The plan I recommended for crossing Harlem River was by a bridge fifty feet (afterwards changed to sixty-five with an arch of one hundred and twenty-feet span) above high water in the river, on which iron pipes were to be laid to conduct the water over the river. The object was economy as compared to a bridge on the grade of the aqueduct. Such a plan in general had been proposed by Dr. John Martineau in his report of 1834. In his report to the New York Water Company in 1826, Canvass White had recommended a similar plan for carrying the proposed Bronx Aqueduct over Harlem River. Noticing these propositions by very respectable engineers, and in view of the economy of the method, and further, noticing there was no navigation on this portion of Harlem River, I had no hesitation in recommending to the Board of Water Commissioners, as a measure of economy, the plan above mentioned.

In December, 1837, I submitted the location and plans of work for the aqueduct on New York Island. This report was approved by the board of commissioners in their report of January 2, 1838.

In May, 1838, the commissioners placed under contract the

portion that remained to the north bank of Harlem River, which included the third and a portion of the fourth division. William Jervis,[4] Mr. Hastie's assistant, was appointed resident engineer and the work directly entered upon by the contractors. This last portion might have been put under contract at an earlier date, except for the legal delays that attended the settlement for land.

In the summer of 1837, I recommended to the board of commissioners the appointment of Horatio Allen[5] as my principal assistant, and he was so appointed to aid me in my general duties.

The work from and including Harlem River bridge to and including the distributing reservoir at Murray Hill[6] was put under contract in October, 1838. This portion of the work had been delayed on account of the embarrassments that arose in obtaining the lands required.

There had been a party opposed to the plan of crossing Harlem River, as adopted by the commissioners. This party increased and through committees of the city council made much resistance. The commissioners, however, were firm and resisted their operations. The point the opponents relied upon was mainly a claim that it would injuriously interfere with the navigation of Harlem River. Had there been an actual navigation, the engineer and commissioners would have provided for it as a necessity; but as no such thing as a navigation that would be interfered with by the proposed plan was in existence, or had hitherto been considered by the public as probable; and as the economy of the plan proposed was important, the board of commissioners regarded it as a duty to resist the efforts for a change. The city council did not sanction the change as proposed by their committee. The advocates of a change to a high bridge had other objects than navigation to urge them. Those owning land near the site believed a high bridge would improve the value of their

4. This is probably Jervis' younger brother, who is known to have been a civil engineer. Another brother, the Reverend Timothy Jervis, was also a civil engineer before entering the Presbyterian ministry.

5. This would indicate a continued congeniality between Jervis and Allen, after the latter's departure to South Carolina eight years before.

6. This famous Egyptian-styled structure was a New York landmark on the site where the New York Public Library now stands.

lands, as giving an ornament to the district. Some citizens of New York regarded a high bridge as a work of art that would be creditable to the city. But all these did not induce the city council to interfere with the proceedings of the commissioners. As the work was only put under contract in October, not much could be done until the following spring. In March, 1839, the party opposed to the plan of the commissioners applied to the legislature for an act to compel them to construct a high bridge. In this they so far succeeded as to compel the bridge to be so constructed that the under side of the arches at the crown should be not less than one-hundred feet above high water in the river, or that the aqueduct should be carried under the river at a depth that would not affect the depth of water in the channel of the river.

This was unsuccessfully opposed by the commissioners, and they had no choice but to obey the mandates and leave the responsibility with the legislature. Personally, neither the commissioners nor the engineer had any choice but to perform the work on the most economical plan. As was natural in such a case, the engineer was charged, as a motive for his plan, that he did not feel competent to the task of building the high bridge, that he wanted the boldness of his predecessor. Of course he did not know what plan his predecessor had proposed, there being no plan left for his guidance, nor was it claimed there was any such plan. It was true he proposed in general a high bridge, without leaving any evidence as to the manner in which the difficulties were to be surmounted or as to what he considered were the essential details of such a structure. On the 7th of May 1839, two days after the passage of the act above referred to, the board of commissioners held a meeting and took steps for the vacation of the contract for section 86 (the bridge section); and also gave instructions to the Chief Engineer to prepare plans and estimates for the two methods of crossing Harlem River proposed by the new law. In accordance with this, the chief engineer submitted plans and estimates in a report of the 8th of June 1839, or one month after receiving instructions. In this report the chief engineer found the tunnel would probably be less expensive; but though the high bridge would cost about two hundred thousand dollars more than the tunnel, he gave what he considered sufficient reasons for preferring the high bridge.

On August 13, 1839, the high bridge was contracted for.

The plans, the methods of constructing the several works as the tunnelling, the conduit, the ventilators, the waste weirs, culverts, foundation, and protection walls, and the excavations, filling and so forth, and form of ground at different places, with the specifications and forms of contracts, will I think be sufficiently explained by the drawings and papers I have sent for deposit to the American Society of Civil Engineers, and may be seen at their office in the city of New York. I shall therefore confine my further remarks to those structures which are of more special importance.

Aqueduct Bridge at Sing Sing

The only peculiarity it is necessary to notice is the plan for cast iron lining for the conduit. This was introduced in order to guard against leakage that by possibility might percolate through the masonry. There was good reason to suppose such would be very small and of no importance as to waste of water; but in this climate, leakage amounting to only a sweating of the arch stone in the bridge masonry would tend to disintegrate the most durable stone, and it was therefore important to arrest it by every means practicable. This iron lining appears in the cross section of the conduit over the bridge. The method has been in the main successful, though it requires some attention to correct the effect of longitudinal contraction and expansion. I would recommend in such cases an iron pipe, put together with faucet and spigot joint, as a more perfect method. Of course it would require a large pipe to carry the water of the aqueduct without extra fall, but it is practicable to cast pipe of sufficient size for the Croton conduit.

The centering [7] of the Sing Sing arch was on the plan adopted

7. "Centering" is the wooden scaffolding and forms on which the arch-ring of stones is laid. A critical time is the removal of centering after the arch is formed, because sometimes the bridge collapses or falls askew if the masonry work has not been set properly.

for the Waterloo Bridge at London, and is an excellent one for such arches. As an evidence of the close joints of the masonry of this arch, I had a level taken before striking the centers, and found the settlement or the arch was less than half an inch. The work in this arch has no superior in comparison with other arches of this size of which this item is recorded. The span of the arch is eighty-eight feet.[8]

The receiving reservoir at 86th Street and the Glendenning Bridge will be sufficiently explained by the plans and specifications in the library of the American Society of Civil Engineers.

DISTRIBUTING RESERVOIR, MURRAY HILL

The leading feature of this is the plan of hollow walls. The retaining walls were required to sustain a pressure of forty feet of water. The natural surface of the ground was about the level of the bottom water line of the reservoir. A solid retaining wall of this height would require to be very thick. A single thick wall has the objection that it requires more time to ripen, or become well set in the cement, and further, the cost of such a wall would be greater for the same support than a wall having a broader base and less aggregate material. In this view, I prepared the plan of a hollow wall, that is, two parallel walls connected by cross walls, and the cross walls connected at the top by brick arches. This kind of work was carried up within about twelve feet of the top or coping of the reservoir, and then a single wall completed the work. An opening was left in the cross walls so that a man may go around the whole and discover any leak there may be in the masonry. The working of this plan has proved very satisfactory.[9]

The general facade or outer wall had a bevel of one in six, and was coped with an Egyptian cornice which was most suitable for the general style of the walls.

At the close of the year 1839, all the works had been put under contract except the grading for the line of pipes on Fifth

8. The preceding extract was selected by Jervis from his *Memoir.*
9. *Ibid.*

Avenue. The works on the Island, having been more recently put under contract, were much behind the works in Westchester County. The Harlem Bridge on the high plan had only been a few months under contract, and very little work had been done at this time.

At the close of 1839, there had been fifty-four sections of the aqueduct completed and accepted—aggregating nearly twenty-one miles. On the remaining part, the masonry of the aqueduct had been constructed for an aggregate length of five and one-third miles. Total a fraction over twenty-six miles of the aqueduct masonry completed, leaving with the lines of iron pipe about fourteen miles in a more or less advanced state to complete the work. It will be noticed the unfinished portion included the reservoirs, Glendenning Bridge, and the Harlem Bridge on the new plan. The total expenditure at this date (1839) was a trifle under four millions of dollars.

Flow of Water in the Aqueduct

After the water was let into the aqueduct, I entered on a series of experiments to ascertain the actual flow of water. In a variety of methods this was very satisfactorily ascertained. At the time of experiments, on actual measurement, the depth of water was two feet. It was conclusively demonstrated the flow was at least twenty-five percent greater than given by the formula on which it had been calculated. It may be noticed the formulae of Prony, Eytelwein, and Robison had been used.[10] These vary somewhat, but not materially. Robison gives a larger flow for larger volumes and a less relative flow for small volumes. This I consider as a result from the larger volumes and slow velocities he used

10. Gaspard Clair François Marie, Baron Riche de Prony, was a famous French hydraulician and director-general of the École Nationale des Ponts et Chaussées. Johann Albert Eytelwein was a German army officer who later engaged in river and harbor work in Prussia. His *Handbuch des Mechanik fester Korper und Hydraulik* (1801) is an engineering classic. John Robison, a friend of James Watt and John Rennie, was professor of mechanical philosophy in Edinburgh. He wrote *A System of Mechanical Philosophy* (Edinburgh, 1804) and contributed to the *Encyclopaedia Britannica*.

in his experiments. All the formulae noticed are safe methods of computation, and for such a work as the Croton Aqueduct it is quite safe to add twenty-five percent to Robison's formula. A full aqueduct would increase this percentage. Similar trial was made on the Boston Aqueduct, where the inclination was less and the flow less rapid, and this case proved the flow to be thirty-three percent over Robison's formula.

I now go back to an incident of the autumn of 1841. A professor of natural philosophy gave a public lecture in Clinton Hall. In this he criticized the strength of the Croton works generally, but especially the iron pipes laid in Manhattan Valley. These he said were entirely inadequate to the strain to which they would be exposed, and he predicted a total failure. At the time of the lecture these pipes were laid, and were expected to be the channel through which the water would flow early the following season. As a matter of course, many very intelligent people took it for granted the learned professor knew what he was talking about. Such a statement could not but have an unfavorable influence, even on the friends of the aqueduct. The criticism of the professor could not disturb any man familiar with such investigations; but it is to be remembered there were comparatively very few men in the community who understood this subject. It could only be expected of these who had examined it as experts. It did not in the least disturb my confidence, and had it not been from the novelty of this class of works in the community, not much attention would have been excited by the criticism. These incidents did disturb the engineers, more especially those claiming there would be a comparatively small supply of water from the aqueduct. It may be asked, "Why did these erroneous criticisms disturb confidence in the commissioners?" Evidently, because they were made by persons assuming or holding positions which gave faith in their statements. To most men they appeared as of professional value, and though the engineer should deny the criticisms, he would be regarded an interested party and at most his denial would be taken merely as a case "where doctors disagree."

I notice these incidents to show a historic fact—that the works of the Croton Aqueduct were subject to a sharp criticism through-out their construction and that several circumstances conspired to make them very annoying to the engineer. I never regarded them as of any importance, except as they impaired the confi-dence of the public and the water commissioners in the engineer-ing of the works.

At this time it is difficult to appreciate the force of those criticisms. If the works materially failed, their expense would entail a heavy debt on the city—a debt at that time regarded a very serious burden. The ordinary expenses of the city govern-ment at that time was about one and a quarter million dollars per year, and a debt of ten millions appeared large. The fear of heavy taxes on account of the aqueduct gave serious anxiety to property owners, even to those who expected a good supply of water. The financial views of the citizens did not then anticipate so early a day when they would pay six millions of dollars for a court house, or that the state would pay fifteen to twenty millions for a state house for its public officers. In these respects the times have greatly changed.

It was manifest the cost of the aqueduct would be large, and there was every inducement to economize when it could be done with safety. All men who have had large experience in expensive works know the danger of inadequacy when there is anxiety in regard to expense. But in view of this there was the intention of omitting no expense regarded as material to safety. For my own part I could not have been more deeply impressed than I was of the grave responsibility I had undertaken, and I think all the different boards of water commissioners felt the same.

IX

CROTON DAM

BEFORE the report above referred to was submitted to the city council, and before I had seen it, a most serious disaster happened to the embankment of the Croton Dam. This was on the 8th of January 1841. The water rose to within a few inches of the top of the embankment, and made a passage between the frozen and unfrozen earth; this occurring in the night, it was impossible to prevent a breach, and the bank was carried away. The water rose about fifteen feet above the lip of the dam, pouring over the masonry an immense volume of water. The masonry was on solid rock and was not disturbed by this severe action. With the president of the board (Samuel Stevens) I immediately proceeded to the scene of disaster. On passing over the hill as the road entered the valley, the view was indeed sad and the aspect was severe in the extreme. No other of the works on the aqueduct were injured. No one without such experience could imagine the severity with which this scene, with its attending circumstances, affected me.

Damage to the work was intensified by the loss of three lives and much destruction to private property on the river below the dam.

Of course this made a new paragraph necessary for the report above noticed. I had nothing to say but to state facts, which I did briefly the evening of the next day in a letter to the board from Sing Sing.

The embankment was gone, and it was manifest I had been in error in regard to the extent of waterway necessary for so great a flood. For three years I had seen no flood that gave me the suspicion of one of this magnitude.

I did not then know how the board of commissioners might be influenced by this event, or in fact the views they had stated in their report which I supposed was about ready to be submitted to the city council. This I did not expect to see until it was published by the city council several weeks later, and remained with such impressions as I could gather from occasional conversations with the commissioners. Though I could not know particularly the effect on the minds of the commissioners, or whether they would sanction the plans I might propose to repair the loss, I immediately gave my mind to the consideration of a proper remedy.

It was not to be doubted that a large addition to the weir of the dam was necessary. This brought me to face a difficulty I originally aimed to avoid. In order to obtain all the weir I could for the old dam, I had occupied all the available rock for foundation. Beyond this to the north or opposite bank of the river, the bed was gravel. To provide works for the overfall of forty feet was a difficult matter. The greatest difficulty in guarding against the overfall for dams in large streams is found to be in providing against the injury of the falling water on the bed of the stream below. Even a bed of solid rock has often been broken up and large masses taken out by the force of the falling water. I had not known of so high a dam resting on a gravel foundation. In this case the works must all be artificial. I could learn from experience of such works of no method that promised the security this situation demanded. No small difficulty appeared in the fact that the work must be constructed in a river exposed to frequent rises or floods.

The idea occurred to me that some plan must be adopted by which the water in its passage from the lip of the dam could be turned gradually from a vertical to a horizontal position by the time it reached the apron below. By this plan, the flow of the falling column of water would move over the apron with the least force to disturb it or the bed of the river beyond.

Nothing but masonry could be regarded proper for the dam. I finally hit upon the plan of forming the lower face of the masonry on an O.G. [ogive] or reversed curve that would carry the water down on a smooth volume from its starting at the lip to the apron below. The tangent of the upper curve was a horizontal line through the lip of the dam, over which the water moved on the curve until it met the tangent of the lower curve, when it began a reverse from its preceding direction and rapidly approached the lower tangent of the lower curve, which it left with a slight angle above a horizontal line. This form was adopted, and has worked very satisfactorily, the water in almost all stages moving on the curved face of the masonry.

This method was very favorable in modifying the form and giving a direction more easily managed to this heavy column of falling water—falling forty feet from a depth of several feet on the lip of the dam; still, in high floods it had a large power and required great precaution in providing an apron to carry it so far as would prevent dangerous action on the gravel bed of the river. From the termination of the masonry an apron was constructed, extending downstream fifty-five feet. This was formed by a timber crib, the first sixteen feet being filled with concrete masonry and the balance with loose stones. On the first thirty-three feet it was covered with six-inch white elm plank and the remainder with hemlock plank. About three hundred feet below this apron a dam was erected of about nine feet in height, to form a pool to break the force of the water after it had left the apron. This worked very well for several years; but on the occurrence of a very high flood, which I am informed rose eight feet on the lip of the dam, some of the apron planking was lifted as was supposed by the reaction of the water below the dam. The planks were replaced with spaces of one to two inches between them, which allowed the reaction a relief between the planks, and so far as I have heard, this has proved a remedy for the reaction. In the flood above mentioned, the secondary dam was carried off.

COFFERDAM AND RIVER WORK

At a point in the edge of the river at ordinary flow of water and about seventy feet from the abutment of the original dam, a cofferdam was constructed of concrete masonry which served to cut off the water of the river from this point to the north shore. One of the pieces of this coffer was sunk in water from six to ten feet deep. By this means the foundation was laid bare from the coffer to the north end of the dam. This gave me means of putting in a foundation and carrying the masonry several feet above the level of low water in the river. After raising this masonry about eight feet, it was then necessary to provide for the portion of masonry across the seventy feet of river channel to the wall of the original abutment on the south side. The depth of water in the river was from six to fifteen feet through which the water of the river passed

There being no way to divert the water of the river at any reasonable expense, it was necessary to do the work in the water. For this purpose two timber cribs were sunk in the river, one above the highest part of the dam, and the other below with a space of sixteen feet between them. The inner faces of these cribs were of hewed timber. The cribs being sunk, the space of sixteen feet was filled with hydraulic concrete masonry. The concrete was put in by means of plank spouts [wooden chutes]. The bottom of the spouts rested at first on the bed of the river and were raised as the concrete filled up. The top of the spouts were of course above the water in the river. This central concrete wall was raised to nearly twenty feet above the surface of the water in the river, making it in greatest height about thirty-five feet.

Above the upstream pier, a filling of earth was put in. A good deal of water found its way through the piers, but as the concrete rose this was checked, and the water gradually raised and formed a pool above the works until it rose six feet to the level of the culvert in the original dam and then had an outlet through the same. There was no escape for the water until it rose as above stated to the culvert in the original dam. By the earth filling as above noticed, the water was shut off from the works until it rose to and was discharged by the culvert. This was a time of extreme anxiety, as any considerable rise in the river would have been very embarrassing. There were several rises that did some damage and impeded the operations. When the water of the river had been turned through the culvert, there was little difficulty in preparing the foundations below the piers or cribs above described.

The work on the apron was now carried forward and brought up to its position as a buttress to the lower curve on the face of the dam. The old culvert had sufficient capacity to carry the ordinary flow of the river and something more. We were therefore protected from the ordinary flow and from light rises in the river. For any considerable rise in the river, we were exposed to a flow of its water over our works.

To aid the culvert in any rise it could not discharge, temporary aprons were constructed to carry the surplus water over. These were placed in the lower parts of the work and changed from time to time as the work advanced. These means provided for considerable floods, but at times they were insufficient, and the water flowed over the works, impeding progress and doing some damage.

As soon as the apron was brought up and the foundation prepared, the masonry abutting against the apron was put down. The first three courses of the face stone in the lower curve were slightly curved and all dovetailed together. These face stones were of granite and were carried up as fast as the masonry was ready to receive them; they were closely cut and formed a good face for the water to pass over when the rise could not otherwise be provided for. When this occurred over parts newly laid, some injury was done to the mortar joint, but the close cutting of the stone prevented serious injury to the face stone.

It was designed to lay the face stone in the curve of the lower side of the masonry as fast as the work was brought to the proper height. But some disappointment occurred. The stones were brought from the eastern states, and in their preparation and transportation such great distances delays had impeded the work and so retarded this portion of the work. At one time, in the early part of autumn, quite a severe rise in the river occurred passing over a large part of the work. The masonry proved sufficient to bear this operation without very serious damage. In the lately laid masonry, the mortar in some parts was washed out, and the stone required to be relaid.

In some respects disappointing, still the work at the close of the season was brought up to about twenty-two feet above low water in the river. A timber aproning was then put over the whole of the masonry to protect it from the winter and spring floods. By this means the water of the river passed over without serious damage.

In view of the magnitude and the difficult character of the work, the progress for one season, though less than required by the contract, evinced a high degree of energy on the part of Messrs. McCulloch, Black, McManus, and Hepborne, the contractors. The contract was made in April, but the state of the water in the river did not allow much work to be done until June. From the middle of June to the 1st of October, the weather was as favorable as could be expected. The flood of the 1st of October arrested the work for a week, and was much more serious than had occurred since June. After the subsidence of the October flood, there was no material interruption of the work until near the season for suspending masonry.

This was a season of intense anxiety to me. I watched the indications of rain very closely and devoted as much personal attention to this as I could possibly spare from other work.

In my report to the commissioners in July, 1842, occurs the following: "In looking back to the times where danger stood before us, I should not do justice to my feelings, did I not acknowledge the hand of that Benign Providence who has so kindly tempered the storm, and saved us from material harm."

On the opening of the spring of 1842, the temporary apron over the masonry was removed as soon as the spring floods had subsided. We were very anxious to carry the masonry up so as to turn the water into the aqueduct as early as practicable. The masonry was now much narrower than that built the previous season, and the contractors resumed the work with energy. After the 21st of May the weather was favorable until the 22nd of June, when the water was turned into the aqueduct, though the masonry was yet ten feet below full height. On June 27, the water reached the receiving reservoir at 86th Street, and on the Fourth of July the distributing reservoir on Murray Hill in New

York. After, the work on the dam was very much embarrassed by floods in the river and was not completed until late in the season of 1842. Had the previous year, when the work was first commenced, been similar to this, very little could have been done. Even with the additional aid in discharging the floods which we had in the aqueduct, it was found very difficult to complete the last ten feet of the masonry.

The report of the board of commissioners of January, 1841, to which I have referred, and which noticed the loss and damage to the Croton Embankment, and to which I naturally looked with solicitude, came out in print while I was intensely occupied with the plans and specifications of the new work. I found it to be a candid statement of the disaster, of which I had no occasion to complain, and was left to perfect my preparations in the expectation I should be allowed, so far as I might be able, to remedy the loss that had occurred. Considering Major Douglass' communication of about two months previous, this was as favorable as I could expect. There was no comparison of my plan with anything that had been proposed by Major Douglass. There was no ground for any such criticism, as the provision for waste on my plan was greater than that suggested by him in his report of February, 1835, and I knew of no other expression by him on this point. Still from the loose way in which comparisons had been made, I had apprehended they would be made in this case. The fact is, we were both in error as to the extent of waste required.

THE BORREN AFFAIR

While I was occupied on the plans and specifications for the extension of the Croton Dam, and had made considerable progress, there came up a new element of excitement. A Mr. Borren, an Englishman claiming to be an engineer, made an adverse criticism on the works of the aqueduct, and especially charging that it would fail to deliver any considerable quantity of water. This Mr. Borren sent to the governor of the state a lengthy paper. The governor sent it to the water commissioners, and they handed it to me to examine. The main point as I have stated was that the quantity of water the aqueduct would furnish would be very small and far below what the engineers claimed. This criticism presented no definite method of computation, and it consequently presented nothing that could be replied to by any recognized principle in hydraulics. Mr. Borren, as I was informed, had several personal interviews with the water commissioners, in which he aimed to satisfy them that his criticism was correct. His paper was examined by myself and two of my assistants, and we had no difficulty in reaching the conclusion that Mr. Borren did not understand the matter he was discussing. But it was evident he made some impression on the minds of the commissioners.

In order to obtain a test of Mr. Borren's ability of hydraulics, we prepared a hypothetical case, giving him all the elements necessary to a calculation, and asking him to determine the quantity of water that would be delivered. The case was given as hypothetical, only leaving Mr. Borren to determine the flow of water from those elements. The case, though presented as hypothetical, was an aqueduct that had for years supplied a city in France with a well ascertained quantity of water. The object was to put Mr. Borren to the necessity of presenting his method of computation, in order to learn if he could make or by what process he reached conclusion that the Croton Aqueduct would fail. As Mr. Borren never called on me, it was necessary the above proposition should be presented by the water commissioners, who readily consented to submit it to him. After keeping the proposition some time, Mr. Borren sent the commissioners a lengthy reply; not made up of specific calculations, but a rambling sort of remark, and finally reaching the conclusion, the case could afford no water. As we knew the facts as to the delivery in the case presented, and also that Mr. Borren failed to show any evidence that he could make such investigations, we saw full confirmation of previous opinions, from Mr. Borren's own pen—that he did not understand, and could not make any legitimate calculation in hydraulics.

We stated this result, and showed the failure to the commissioners. The latter could not gainsay, neither could they appreciate the position fully. But it weakened Mr. Borren's influence. Mr. Borren had made some suggestions as to the plan of works for the extension of the Croton Dam. On this point there was another Englishman (I do not recollect his name) who I saw several times, and seemed much disposed to give advice as to the manner of extending the Croton Dam. He made various suggestions on the general works of the aqueduct, but mostly in reference to the Dam. He was quite desultory, and I was unable to see what his plan was. In order to get at his views, I proposed to have a meeting at the residence of one of the commissioners (John D. Ward) with the English engineer, to be accompanied with two of my assistants, Messrs. H. Allen and P. Hastie. Commissioner Ward was a mechanic and I judged would more readily appreciate any mechanical principle that might be suggested than any of the other commissioners. Accordingly the meeting was held; all the parties came together at Mr. Ward's house. Conversation began, and it was stated by me that the object was to ascertain what our English friend would propose as a plan for the extension of the Croton Dam. He commenced by discussing the merits of Portland cement, and other matters equally irrelevant to the object of the meeting. After waiting some time, I referred to the specific object, and requested his attention; but he did not seem inclined to be specific, and as the evening was wearing away, I again called his attention, and requested him to state his views of such plan as he would recommend; that the evening was passing away, and the object of the meeting would be lost if we consumed the time on other objects. He then stated as an outline what he would recommend. Of course I had nothing to say. I asked a few questions to give him opportunity to explain his plan more fully, so that we could see what he intended. Of course, he could only present a skeleton or general features. I looked at Mr. Ward to judge the impression made on his mind by this development; but he like a true judge, gave no reading in his countenance. Soon after, the meeting broke up. After myself and assistants left, we could only indulge in surprise that a man after such pretensions should exhibit such total lack of engineering capacity. He did not present a single feature that could be regarded as worthy of an engineer.

I never learned what opinion Commissioner Ward formed of the interview. A few days after, the chairman of the board, Samuel Stevens, had a conversation with me, in which he alluded to the English engineer in a complimentary way. I had been very cautious, but this after the interview as above was rather too much, and I quietly replied that I had very little respect for their opinions.

Of course such incidents consumed time and were calculated to delay progress. But the plans and specifications were pushed forward. The situation not only involved the kind and manner of work, but also the order of construction so as to provide as far as possible against the contingency of sudden rises in the river.

After what I have related above, I completed my plans and specifications and submitted them to the board of commissioners, explaining all matters to them so far as they desired, and they approved and ordered the work advertised for contract.

Although I have presented the men in the above remarks as claiming to be English engineers, I have no idea they had any standing as such in their own country. They no doubt were of a class that supposed it was not possible anyone should know much of engineering in the woods of America. It may be asked —How then could they influence intelligent Americans? For the reason that the subject was at the time a novel one, and those not engineers were in doubt, and looked on the question as one "where doctors disagree." At the same time it was obvious that many intelligent men, not engineers, looked on the matter with discrimination, and so steadied public opinion that the work was not arrested. Among these, I regarded John D. Ward, one of the board of commissioners, a man much respected by his associate commissioners. He was a man of few words, but I have always thought his influence was favorable in sustaining the order and advice of the chief engineer. The result is now known, and I only refer to these embarrassments to show the perplexities that pretentious men may throw in the way of a public work not familiar to the citizens.

X

HIGH BRIDGE

IT HAS BEEN STATED the High Bridge was put under contract in August, 1839. I have before stated the reasons why I had recommended the low-bridge plan. It was natural that an engineer should incline to a work that would give prominence to professional character as a work of art. In this I was ruled by the duty of effecting the improvement by the most economical method practicable. The Democratic and the Whig boards of commissioners did all in their powers to sustain my views, but the question was authoritatively changed as before stated, and plans for the High Bridge were ordered, and the work put under contract. It now became important that the matter previously looked at in a general way should be more thoroughly understood.

It had been supposed a rock foundation would be found for the piers of the bridge. Rock in places was found on each side of the river, and though the soundings in the river had not in all cases met rock, it was supposed it would be found within limits that could be reached. But more thorough examination failed to show rock in some places after going eighty feet below high water. What was originally supposed in some cases to be rock in place proved to be only large boulders that lay very thick in the mud and sand, and below these a bed of sharp sand.[1] These boulders were a serious impediment to the aligning of the cofferdams, and to driving the sheeting of the coffer after it was sunk. As far as practicable the mud was dredged out, and many

1. The problem of discriminating between boulders and bedrock is even today often a difficult decision for the trained foundation investigator.

of the boulders of two to six tons weight were lewised [2] and taken out in ten feet of water.

It was now evident that pile foundations would be required in several piers. If all had required piles, the difficulty or uncertainty as to settling would have been less.[3] Rock was found in place for a part of the piers, and to guard against unequal settlement the pile foundation must be made equally secure against settling.

The first question to be settled was the weight that could be safely rested on the piles without yielding. I had known a good many cases where bearing piles had been successfully used, but these were for much less weights, and I could not find any specific experiments that warranted full confidence for this bridge. I therefore proposed an experiment to determine the question.[4]

I placed the experiment in charge of Horatio Allen—principal assistant engineer. Mr. Allen had a hydraulic press constructed and so arranged as to bring the pressure to bear on the head of the pile. The operation was conducted by Mr. Allen with great care, giving great confidence in the experiment. The trial was made on four piles. I give below, for the benefit of those who may be interested in such questions, the experiment on one of the piles.

This pile was round oak, fifteen feet long, and had been driven about twelve feet fifteen inches diameter at the upper end and probably eleven inches at the lower end. A piling machine was placed over it before the trial, and three blows struck with a hammer weighing fourteen hundred pounds, falling through a lead of thirty feet. The pile sunk under the hammer an average of one inch at each blow. A lever was fitted to determine the motion of the pile under the pressure of the hy-

2. "Lewising" is a procedure for removing large buried boulders by drilling a hole in an exposed portion, inserting a metal "plug" to which a cable is attached, and hoisting out the boulder by a winch.

3. The foundation engineer of today is still confronted with much uncertainty when attempting to predict the differential settlement between piers founded on rock or those on piles.

4. This experiment may well have been the first full-scale test of pile foundations in the United States.

draulic ram of one to fourteen. This pile withstood a pressure of sixty tons without any perceptible movement. It yielded a trifle under sixty-five tons but resumed its original position after the pressure was removed. This yielding was doubtless due to elasticity in the pile and in the medium in which it was driven.[5]

This experiment gave a satisfactory basis for calculating the strength of bearing piles to sustain a superstructure. In this work it was important, as the piers were high and narrow and had to carry the weight of large arches with spandrels and parapets.[6]

The bridge consists of seven arches, each of fifty-feet span, and eight arches of eighty-feet span. A short pier of solid wall at each end completes the bridge. The width of piers at the spring line for the fifty-feet arches is seven feet, and for all but three of the large arches, fourteen feet. The other piers were twenty feet in width as piers of equilibrium. The width of the bridge on top of the parapets is twenty-one feet. The piers and the general façade of the bridge is on a level of one in forty-eight.

Of the land piers, one is on rock on the north side of the river, and one on rock on the south side of the river, as also the abutments. The remaining five land piers are on piles.

When the piling had been driven for the land piers, the earth was excavated to the depth of three feet below the heads of the piles, and the whole space filled around them with hydraulic cement to the top. Then a course of dressed stone, each stone laid on the top of a pile (to avoid the use of timber), and the spaces (a few inches) between the stone filled with concrete masonry. The next course was properly bonded on this. By this course the heads of the piles were excluded from air and were so far below the surface that no material change of temperature would affect their durability.[7] The cement in a short time would

5. This interpretation of the experiment is in keeping with modern theories which were not widely applied until almost a century later. The experimental procedures themselves are very similar to present practice.

6. "Spandrels" are the spaces between the arch ring and the roadway (waterway, in this case); "parapets" are the portions of the sidewalls which extend above the roadway.

7. These piles have satisfactorily performed their function for over a century and a quarter.

become hard and solid and make a good foundation of itself.

The first water pier on the north side was founded on piles— the second and the third on rock—the fourth, fifth, sixth, and seventh on piles—the eighth on rock. The difficulty of adjusting the foundations for a structure of such height and weight demanded the most careful consideration in providing for an even support for the whole structure. The piles were mostly of oak timber and driven from thirty to forty-five feet. The heads of the piles were capped with white oak timber and the spaces filled to the top of the timbers with concrete masonry. Then a similar course of timber was laid across the first course vertically over the heads of piles in the first course, and the spaces filled with concrete to the surface. On the last course, oak planking nine inches thick was laid on which the stone work was commenced.

Each course of stone for the piers, as also the arch stone, were all got out on a plan made for each course, and nothing was left to the judgment of the workmen as to breaking joints. This secured the most perfect bond in the masonry. The river piers above high water line in the river were made hollow, only having such an amount of material as was required for strength. This method counteracts any tendency to unequal strain. It had been adopted in the heavy class of English bridges with very satisfactory results. Passages were provided in the spandrel walls to lead any water which might fall between the parapets into the hollow spaces in the piers, and so carry it down to an opening in the piers near high water line in the river. The obstacle presented by the large boulders that were imbedded in the mud and sand has been noticed in general. I insert below an extract from my report to the commissioners [for December, 1843] after these difficulties had been surmounted.

The obstacles to the work of cofferdamming arising from the stratum of boulders running across the river, at a level that did not admit of their removal before the coffer was placed, were very serious. The boulders lay so close and compact that the sheet piling was to a great extent rendered useless after reaching them. Yellow pine sheeting timbers, nine to twelve inches thick, were to a great extent driven to splinters or broken on reaching the boulders, rendering them of little use for the work required below the surface of the boulder stratum. The excavation necessary in removing the boulders, and the gravel and sand intermixed with them, having very little support from the sheet-piling, was attended with much difficulty before the inner sheeting of the supporting timbers could be brought to effectually resist the earth at the sides and prevent it from running in from beneath and behind the coffer. In several instances the coffers settled considerably before the lower work could be secured and gave serious apprehension for the safety of the work; but notwithstanding the severity of the trial, they proved sufficient for the pressure until the underwork was carried down to the requisite depth and secured by sheeting supported by heavy timber struts. This work has been carried forward steadily, but not hurried. No serious disappointment has occurred in its progress, and though numerous delays have been experienced, they have been as few as could be expected in prosecuting such a work. For two of the river piers, numbers 8 and 9 (regular series being the second and third water piers), rock foundation was obtained at a depth below the top of the cofferdam (which is one foot above extreme high water) of twenty-one and one half feet for No. 8 and thirty-nine and one half feet for No. 9. For the remaining five river piers no rock could be found, and piling has been adopted. In consequence of the size and compactness of the mass of boulders above mentioned, it was necessary to excavate and remove them before a proper medium could be reached for driving bearing piles, which required the pits to be sunk within the coffers before the piling could be properly done to the required depth (twenty to forty feet below high water), The bearing piles have generally been driven from thirty to forty-five feet. Great care has been taken in driving the piles, adapting the length to the medium in which they were driven, and when the long ones did not appear fully satisfactory, intermediate piles were driven until the resistance indicated adequate stability to sustain the structure.

That I may not seem to give undue prominence to this work, I make the following quotation from the report of the board of commissioners (Messrs. Samuel Stevens, John D. Ward, and their associates) of 11th of January 1841. In speaking of the Harlem

(High) Bridge, "It is a fact not to be disguised that the erection of this bridge is not only a *stupendous* but is a *Herculean* task for our city to execute, and requires more engineering talent, inspection, and watchfulness than any other part, or we might almost say, all the other parts of the aqueduct put together."

The commissioners had frequent opportunity of inspecting the work and were impressed with its difficulties; but the following season's operations were attended with greater embarrassments than had previously been met.

The carpenter employed by the contractors, a Mr. Bishop, proved to be a very valuable man. He invented and erected a derrick that was worked by steam power, and readily took up any material or machinery and placed it on the bottom of the pits or on the top of the parapets. In raising the centers for the arches, he took up the largest timbers, raising them a hundred feet and placing them in position with great ease and expedition. The stone were nearly all laid by this kind of derrick.

As before observed, it was designed to place between the parapets on the top of the bridge two cast iron pipes, each four feet in diameter. These pipes were capable of conducting the full flow of water as the masonry conduit of the aqueduct. This was the full service for which they were wanted.

The space between the parapets was sufficient for the pipes. The plan of two pipes instead of one of large caliber was preferred for the reason that the pipes of iron could not be considered indestructible, and also, they would be liable to accidental injury. Repairs or removals would inevitably be called for at some time, and as they were fourteen hundred and seventy-five feet in length, considerable time would be required espeically for removal. If there were two pipes, one line would be left for the supply of the city while the other was in process of repair. In important works, it is regarded as sound policy, so far as may be practicable, to make such parts as are liable to fail through accident or decay in such a way as to admit of repair or renewal, without suspending the usefulness of the work. In all cases where iron pipes were laid, this dual principle was adopted. The receiving and distributing reservoirs had this provision.

The works for supplying a large city with a daily necessity is eminently of this character. So much was I impressed with the importance of this view that I made estimates to ascertain the extra expense of constructing two masonry conduits over one of large size. I found the excess of cost to be such as to present a financial objection that could not at that time be overcome. There was no such difference in regard to the pipes over this bridge, and as the pipes had no such promise of durability as the masonry conduit, the duplicate arrangement was clearly called for. The plan was simple and extremely appropriate; nothing could be more so. The capacity of the aqueduct was regarded as making provision for many years in the future. In this view, it occurred to the water commissioners a saving of present expense could be obtained by substituting three- instead of four-foot pipe; that the three-foot pipe would probably supply the city until a renewal of pipe would become necessary or so long as to make it a matter of economy to put them down in the first place. In this view the three-foot pipes were put down. In the meantime, the works of the bridge and the gate houses were adapted to receive four-foot pipes when they should be found necessary. The temporary line of pipe laid down in the valley to supply the city while the work of the bridge was unfinished was taken up as soon as the line of conduit had been connected over the bridge. These pipes were readily taken up, and as soon as one line over the bridge was brought into use, they were used for the second line, showing there was no difficulty in such an operation. At the time, these three-foot pipes abundantly supplied the city.

The consumption of water in the city increased far beyond the original anticipations, and the demand for the four-foot pipes came much sooner than was expected. When it came, there was a most simple remedy, namely, to take up one of the three-foot pipes and put down a four-foot pipe, for which every provision was made.

About the year 1852, or ten years after the water was brought to the city, there began to be a call for more water than the two three-foot pipes could furnish. I discussed the subject several times and fully explained to the engineer, Mr. Craven,[8] the plan and the obvious course to be pursued. Without explaining to me, it appeared from his remarks that he was desirous of a different plan, but said no plan should be adopted without consultation with me. This proposition he did not carry out.

Instead of this simple process, one has been adopted that wholly changes the simplicity of the original design. Instead of taking up the three-foot lines, they are both left, and an opening made on the side of the gate chambers into which is inserted a wrought iron pipe. The latter was carried round to the pipe channel of the bridge and placed over the two old three-foot lines, resting just above them. The parapets of the bridge were raised and a brick arch thrown over the pipes. This arch had for its abutments the parapet walls of the bridge, and being rather flat, it was deemed necessary to put in iron rods to guard against the thrust of the arch against the parapets; these rods ran across from side to side through the parapets of the bridge.

The necessity for repair or the renewal of the pipes or of the iron rods does not seem to have been provided for. The pipes and the stay rods will surely go to decay and some day repairs will be necessary. When this contingency arises with respect to the stay rods, new ones may be put in, but if the failure is not noticed in time, the parapets of the bridge may be found out of line and seriously injured. In regard to the pipes, when renewal becomes necessary, the old three-foot lines if still sound may serve the city while the wrought iron pipe is renewed. This renewal will be materially impeded by the arch over it, and will probably take much more time than would be required to take up and relay on the original plan, a cast iron pipe of the dimensions required. When it becomes necessary to take up the old lines of three-foot pipe, now lying immediately under the

8. Alfred Wingate Craven entered Yale at the age of thirteen, but later switched to Columbia. His early career was in railroads, but he was made chief engineer of the Croton Aqueduct Department in 1849. It was in his office, on November 5, 1852, that the American Society of Civil Engineers was founded. He remained chief engineer until 1868 when he resigned because of poor health. He later practiced as a consultant, but again his health failed and he died at Chiswick, England, where he had gone for an extended rest.

wrought iron pipe, the difficulty will be much greater than to have done this as provided for under the original design. It is manifest the new plan has complicated the facility for removal, giving complexity and expense, instead of simplicity, convenience, and economy in renewal.

I make the above remarks that it may be understood, when the necessity for renewal occurs, that this recent device was no part of the original design, and that I may be exonerated from any censure for the extra difficulties and expense that will be encountered.[9]

It is now very evident, it would have been better economy if the water commissioners had allowed the four-feet pipes to be put down at first, as well as thus maintaining the simplicity of the original design. But the commissioners had no idea the time was so near when they would be required to supply the water demanded. In view of the quantity of water needed, as determined by the experience of large cities, it was quite a prudential measure.

The Harlem Bridge has now stood the service over forty years, and so far as I have learned, no failure has been manifested.

Further disturbing criticisms came out in the season of 1841. The first I notice was on the occasion of gentlemen sitting on the deck of the steamboat passing down the Hudson River. One of these was a professor of mathematics in a neighboring college. The works of the aqueduct were occasionally seen from the steamboat, and as there were some New Yorkers in the party, their attention was attracted to the aqueduct. Naturally supposing a professor of mathematics would understand the theory of such an enterprise, some one asked him what he thought of the

9. Jervis was quite correct in his prediction. Craven's expanded system did crack and leak under increased demands, so the New Croton Aqueduct was constructed (1885–91). It is not known exactly when Jervis wrote this chapter, but according to Desmond Fitz Gerald, president of the American Society of Civil Engineers, "In December 1881, he [Jervis] was summoned to New York by Isaac Newton and E. S. Chesbrough in consultation on plans for the new Croton Aqueduct" (*Transactions*, ASCE, XLI (June, 1899), 612). Jervis was then eighty-six years of age.

Croton Aqueduct. He gravely replied that it was a great work, but added, "there would be great disappointment in the quantity of water that would be furnished—that it would be much less than anticipated." This conversation was soon circulated about the City of New York, and made considerable impression on the minds of men who in other respects were intelligent. Coming as it did a few months after the criticism of Mr. Borren (as above stated) as to the flow of water, it made considerable impression on the water commissioners.

There had already been many persons in the city who were disturbed at the large cost of the aqueduct, and some proposed among their friends to sell out their real estate to avoid the losses on their property from this danger of taxation. It was currently reported of one large holder of real estate that he said to his friends he would sell his property, but concluded he would first have a consultation with John J. Astor, whose opinion on real estate he highly valued. Accordingly he visited Mr. Astor at his summer residence at Astoria. After opening his fears to Mr. Astor, Mr. Astor replied very coolly, "we can hardly spend too much for a good supply of good water." It was reported Mr. Astor's cool treatment of the subject induced the property holder to hold on to his real estate. I mention this latter incident as evidence that there were some firm supporters of the aqueduct project.

This incident of the professor of mathematics and the anxiety of holders of real estate so impressed the minds of the commissioners as to lead to a very serious consultation between them and myself. I stated to the commissioners that I had no doubts on the subject and that my opinion was sustained by my principal assistant, Mr. H. Allen, and also by Mr. P. Hastie, the resident engineer of the fourth division, both of whom were quite able to conduct such investigations. That I was satisfied, not only from the theoretical basis of the computations, but also from my experience in such matters in providing feeders for canals when the actual result showed a larger flow than called for by the formula; and that I confidently believed the aqueduct would furnish a larger volume of water than had been promised.

XI

DOUGLASS CONTROVERSY

THERE WAS a good deal of criticism on the part of the public in relation to the character of the work and as to the prospect of success [of the Croton project]. The small amount of experience in this country of this kind of engineering very naturally led even intelligent men to feel there was much contingency in the enterprise. These criticisms found space in the public papers and caused uneasiness in the minds of the friends of the work. In some cases the water commissioners requested me to reply to the public articles in order to allay the impression they made on the friends of the aqueduct. I do not think it worthwhile to notice the particulars of the criticisms as I regarded them of no serious character and only of importance as they disturbed the public mind at that time. They did not, however, appear of such importance as to induce the water commissioners to relax a vigorous prosecution or to make any changes in the order or plans of the work. Sometimes it was alleged the work was insufficient and would not be equal to the exposure. In other cases it was boldly claimed the aqueduct would not fulfill its office, or only in a partial degree. Then complaint was urged that the work was too expensive and would load the city with a heavy and unnecessary debt. So far as related to the cost of the work, there was no ground to apprehend that it would exceed the estimate I made in 1837. The contracts that were from time to time completed and settled showed the cost did not materially differ from the estimate. This later complaint did not disturb those who were familiar with the operations of construction. In a great work and especially one novel in the experience of the community that undertakes it, all criticisms will have an influence on the public, and it will often be difficult to make satisfactory explanations until it may be demonstrated by completion, and this must be waited for several years.

I was not indifferent to those criticisms, for though my own judgment and confidence were unimpaired, they affected to some extent the power that provided the means for carrying the work forward. The criticisms in relation to Harlem Bridge were more personal than any other up to this time. They were intensified by the relation of my predecessor, and some statements were made that were not true. I was satisfied that time would vindicate my proceedings as also those of the board of water commissioners. We were overruled by the act of the legislature, and when it was decided to build the high bridge, it was steadily and satisfactorily proceeded with.

The state elections in the autumn of 1838 resulted in a change of political parties, the Whig party having carried the election.[1] At the ensuing session of the legislature, which convened in January, 1840, the question of a change in the board of water commissioners was discussed and resulted in the removal of the old board and the appointment of a new one about the 1st of March. The new board consisted of Samuel Stevens, John D. Ward, Zebedee Ring, R. Birdsall, and Samuel R. Childs.

In the old board, the appointments in the engineer department were made on the nomination of the chief engineer, with the exception of inspectors of masonry, who were mostly proposed by two of the commissioners who had been extensive builders and well acquainted with that class of men. But even the inspectors of masonry were placed under the control of the chief engineer who had power to dismiss if he thought necessary. The organization of the engineer department was full at the time of the change in the board of commissioners. The new board had a right to change it in any respect they should think proper.

The contracts were made for nearly all the work of the

1. William H. Seward, the first Whig governor, was elected in 1838, defeating William L. Marcy, a Democrat who had been in office since 1832. Jervis' old friend William C. Bouck, a Democrat, defeated Seward in 1842. Seward later gained national fame as Lincoln's secretary of state.

aqueduct, and was in such forwardness that in about two years it was supposed that it would be so far complete as to allow the water to be introduced into the city. There was not much probability the contracts would be disturbed. As to the engineer department, the course of the new board could not be anticipated.

It was soon manifest the new board supposed some of the doings of their predecessors savored of improper political influence. They came into the office and called for information, especially as to the method of giving contracts. The books and papers relating to this subject were in the office of the chief engineer. These had been kept in a systematic way, and I had no difficulty in furnishing all the facts from the books and files of the office. They embraced all the propositions for contracts with detail as to work and prices. It was easy to learn from these who received the contracts, as well as all the unsuccessful proposals. The practice had been for the old commissioners, after the time for receiving proposals had expired, to hand them all over, before opening, to the chief engineer who opened them and made the computations necessary to decide which was the lowest, or what the aggregate of each proposition was, and to hand the list to the board, who then decided on the most favorable. Those books showing each proposal gave the new board a full opportunity to judge of the character of the proceedings. The new board spent considerable time in these examinations, and I did not learn they discovered any wrong.

In the course of a month or two, I learned that one of the new board had canvassed the contractors and engineers and in this made enquiries of them to ascertain if any partiality had been exercised by the old board, more especially in relation to political influence. Before this I had not known the politics of many of the contractors or engineers. The result of the canvass showed that the enigneers and contractors were about equally divided in politics. I did not learn that even one contractor complained of political preference. It was impossible the late board could have aimed at any political favor without my knowledge, and I am quite sure they had no regard to such influence. On my presenting candidates for appointments in the engineer department, there was in no case, by myself or the board, the least regard or enquiry on this subject. I had presented a candidate for principal assistant engineer, whom I had long known as a decided Whig; to which the board made no inquiry except as to his fitness for the engineering duties for which he was wanted. It was the same as to the contractors, very few of whom were personally known to the commissioners, though I was acquainted with most of them.

I was not surprised at the result of the inquiry, as above stated, for though I had not previously known the politics of many of the persons employed on the aqueduct, I knew well there was no political favor exercised or in any way encouraged by the late board. In addition I may say, I have never known a board of commissioners or of directors who conducted the trust committed to them with greater assiduity and fidelity to the trust they had in charge than the late board of water commissioners. I am thus particular for the reason they were charged, and some influential men believed they were actuated by political motives in conducting the work. It was publicly alleged that my own engagement was the result of influence from the Albany Regency.[2] Sure I am no man suggested to me the most distant allusion to such engagement until a committee of the board (Messrs. Stephen Allen and Saul Alley) called on me at Albany and made the proposition. I cannot say what the Albany Regency did, but it would be very strange that such a measure should be consummated without my having some knowledge of it. In no way had I ever attempted to make politics a basis or means of occupation, and no politician has reason to expect this sort of action from me. Some of the reports about this time I knew to be very absurd.

It was very natural for me to suppose the same influence that changed the commissioners might change the chief engineer. I

2. The Albany Regency was an unofficial alliance of Democratic politicians who from about 1820 to 1854 maintained a system of patronage which permitted it to have substantial control of the state. Members of this group included Martin Van Buren, William L. Marcy, and Silas Wright.

was conscious of having faithfully done my duty, and made up my decision to quietly abide the result. I gave the new board freely every information in my power in regard to the work. I had nothing to conceal. As soon as the season opened for masonry, the new board adopted the rule of the old, to make fortnightly visits over the line. I attended them on these visits, giving the same attention as formerly—discussing the work in all its aspects and giving no attention to the peculiarity of my own situation. I made not the least effort to obtain influence to assure my position. Rumor had it sometimes Major Douglass was to be reinstated, and at other times some other name was suggested. It was hinted to me "your place is very much wanted." It pleased God to give me the prudence to hold myself perfectly quiet in regard to these circumstances and to apply myself closely to my engineering duties.

Every day I was becoming more acquainted with the new board and they with me. I soon thought I saw in them a practical sagacity that would not allow them to do any very absurd thing, and I came to have great respect for some members of the board, especially Mr. John D. Ward.

The summer passed along and the prospect of changing the chief engineer seemed to abate. The commissioners seemed more impressed with the magnitude of the work, and though my intercourse with them was necessarily more formal than with the old board, it was courteous and as pleasant as I could expect. Now and then an article appeared, sometimes in the *New York American,* and at others the *New York Courier and Enquirer* which hinted at the necessity of attending to the claims of Major Douglass. At times hints were made to indicate something due to him. But nothing serious was developed or anything offering a contest until a letter of Major Douglass' appeared in the *Courier and Enquirer* of the 28th of October 1840. This was a long letter addressed to the board of water commissioners, and was designed to show that he, Major Douglass, had been improperly removed from the position of chief engineer, and that great failure and damage would result to the aqueduct by the method being pursued. At the same time he said he was ready to do all in his power to retrieve as far as practicable the errors in progress if the commissioners desired it.

The letter of Major Douglass was a criticism on the late board, although mostly directed against Stephen Allen, the chairman of the board, and Mr. Saul Alley.[3] As the letter in some points reflected on my integrity, I sent a note to the *Courier and Enquirer* dated the 30th of October and published by them on the 3rd of November proximo. In this note I stated the points that referred to myself as being in no way true. As the letter was chiefly an attack on the late board, as I have stated, the chairman, Stephen Allen Esq., replied to Major Douglass in the *Courier and Enquirer* of the 12th of November 1840. In his reply, Mr. Allen denied the statements made by Major Douglass and challenged him to the proof. It was now essentially a question of veracity between two very respectable men. I never learned that Major Douglass ever offered to sustain his charges by anything more than his own assertion. As a man of probity, Mr. Allen was highly respected, and his reply was so largely accepted that nothing further was done, so far as I know, with Major Douglass' letter. My own relation to the controversy with Major Douglass I had at a subsequent time occasion to discuss, and which was published in the *New York Evening Post,* and which I propose to introduce at the proper place in this memoir [see p. 165]. In former pages I have been careful to give Major Douglass the credit of everything he has done, so far as I have been able to discover it.

The season (October 26th) was so nearly through when Major Douglass published his letter, it did not allow any change for the current year, and so we passed along without material disturbance.

The new board of commissioners made their first regular report to the common council on the 11th of January 1841. In

3. Douglass asserted that Allen and Alley had accused him of being "partisan in opposition to the water commissioners and that Allen had treated him with 'violence and overbearing conduct.'" After this exchange in the newspapers, Douglass left to accept the presidency of Kenyon College, in Ohio. He was apparently forced to resign this position in 1845 over a quarrel with the trustees. He was later a professor of mathematics at Hobart College in New York State.

that report, they referred to the engineer department rather favorably, and made no suggestion of any change in the department. They made no important criticism of their predecessors. They questioned the wisdom of the street arches in Glendenning Valley, and abandoned three out of six. They reviewed the Harlem Bridge question and advocated a change to the old plan of the predecessors, advising the common council to apply to the legislature for a repeal of the act requiring the high bridge. This was of course a compliment to their predecessors. But, to a considerable extent the public mind favored the high bridge, as I think for the magnificence of the work, and the proposition did not succeed.

Knowing the history of the work as I did, this first report of the new board was in my judgment very creditable to their candor and presented little the old board could complain of.

It has been stated water was let into the distributing reservoir early in July, 1842. That is forty-one years ago. With a small amount of special repairs, it has furnished the city with a regular and far more abundant supply of water than was originally contemplated. It is now said defects have been discovered. It might be pertinent to ask what work of the extent and difficulties of this has for so long a time performed its functions with so small an amount of special repairs? Perhaps I might be excused, after so long a demonstration of the stability of the works and the power of meeting the object of construction, from noticing further criticism. But as my life has been spared to see the criticisms made recently, I will briefly notice one formally made.

This was an address by George B. Butler to the Municipal Society of New York. The society ordered the paper published and referred it for special action to the 19th of March 1879. Some friend sent me a copy.

On page 4, Mr. Butler says, "Major Douglass, the *projecting* engineer, and Mr. Jervis, who executed the work." On page 12, he says, "the rubble wall adopted by Mr. Jervis for its support in crossing ravines and streams, a mode different from that recommended by Major Douglass, as will appear from the plans annexed." Referring to Major Douglass' plan he presents a diagram which is taken from Major Douglass' report of February, 1835, and is there marked No. 8. That as Mr. Butler says, "showing a foundation wall of great width, while the other plan is the one adopted by Mr. Jervis and which has proved insufficient, entailing enormous expense for annual repairs." As to "enormous expense for annual repairs," I have not been able to detect this in current official reports, unless it refers to the expense of taking down the original aqueduct between Manhattan and the receiving reservoir, about two miles, and substituting iron mains at an expense of about four million dollars. Evidently, this is not repairs but destruction.

Mr. Butler's language in the comparison obviously implies that if Major Douglass' plan of embankment had been adopted, it would have been a protection against the failure of Mr. Jervis' plan. Any other view would leave the criticism without significance. As the plan of Major Douglass was not tested, Mr. Butler does not know that it would have been superior to mine. But as it is brought forward under claim of superiority, it is quite proper to examine the two plans. The diagrams presented by Mr. Butler show the Jervis plan to be a regular wall of stone, laid in courses —that of Major Douglass does not show any wall at all. It represents the foundation of the conduit as resting on an embankment of the same material throughout. Mr. Butler, speaking of Major Douglass' plan, says, "showing a foundation of rubble wall of great width." Now, the plan he refers to, No. 8, does not indicate anything that can be called a wall. But he says it was of "great width." It was just the width of his embankment, and has nothing to show it was anything but an embankment. In his report of February, 1835, Major Douglass gives his view of this as follows, "immediately under the line of conduit, a mound of solid stone . . . as the writer has had occasion to experience in similar situations a cheap and very safe foundation. The residue of the embankment after the conduit is built is then to be formed to the necessary height and width good gravel or loam."

It is perfectly clear that Major Douglass did not regard this a wall, but as a mound of stone to be equal the width of the conduit. His use of the term "solid stone" could not mean anything more than an embankment or mound of loose stone put together by the ordinary process of dumping in, as is usual in forming banks of this material. This view is sustained by Major Douglass' estimate for this kind of embankment, which he places at fifty cents per cubic yard on all the sections except the three last, which are on the Island. This price would be wholly absorbed in getting the materials, and nothing would be left for anything like wall. It is simply an embankment of loose stone.

My plan, as may be seen by the diagram Mr. Butler has exhibited, shows a regular wall laid up in courses. I do not claim the wall was as regular in size of stone as Mr. Butler represents it. The wall was constructed by laying over a course of large stone, as they came from the quarry. This would leave many interstices, and these were filled (not by the usual course of leveling) by stone broken on the wall and driven into every space that would take a piece of two inches. These broken stone were leveled up and prepared for the bed of the next course, and so on alternately to the top. If Mr. Butler had not presented a diagram of my wall, it might have been inferred it was inferior to the Douglass plan; but as the two are presented in diagrams, no engineer could hesitate in pronouncing the Jervis wall superior to the Douglass wall for solidity, the property needed in this situation. Now if the Jervis wall had been found insufficient, the Douglass plan was less efficient.

I do not claim that my plan fully met the requirements of the case. It has been tried and that of Major Douglass has not been tried and can only be judged by analogy, and that leaves my plan decidedly superior for the situation. If mine has prove inefficient, it is clear that Major Douglass' plan would have proved more so.

Mr. Butler says, page 12, "the rubble wall was about five miles." Looking over the reports that show the quantity of foundation wall, I am quite confident there is not over half this length. A considerable portion of this was on New York Island, most of which has been taken down and iron mains substituted for about two miles of the aqueduct as before noticed. The most of this class of wall remaining is in Westchester County, and may be two and a quarter miles in total length. I have shown my plan to be superior to the one claimed for Major Douglass; but I do not claim this as making my plan as solid as the exigencies of the case requires. All I claim is that Mr. Butler's comparisons do not support any superiority of Major Douglass' plan over mine. It is not equal to mine as judged by Mr. Butler's diagrams; and when illustrated by what Major Douglass says of his plan, the whole criticism falls to the ground.

I have certainly no desire to criticize Major Douglass' plan and have only gone into this exposé of one item to correct erroneous criticism on a comparison that did me injustice.

I do not claim that I have not erred in regard to the foundation wall. I supposed a well laid dry wall would give adequate support to the conduit. It has supported it with very small repairs for forty-one years and therefore cannot have been grossly in error. It has given no warning that it was not fully sufficient, yet, except in one place (Bathgate's Meadow), no great expense has been incurred for these forty-one years to correct the error.

The error was in adopting a dry wall. It should have been hydraulic masonry. I find by reports of December, 1840, there had been laid of foundation wall about one-hundred and ten thousand cubic yards. This could have been built of solid hydraulic masonry for about five-hundred thousand dollars extra. It was to save this expense that the dry wall was adopted. It is clear enough now that it should have been hydraulic.

The question now is, how shall this be remedied? No doubt different methods may be adopted according to the situation. It is probable there is not over one and one half miles that will require very extraordinary expense. It is certainly wrong to regard the difficulty as irreparable. Probably twenty-five percent of the amount paid to destroy two miles of aqueduct and substitute iron mains on New York Island would remedy the defects complained of and give the aqueduct great permanence. To complain for years of defects in the aqueduct and give no vigorous attention to repairs is no part of wisdom in maintaining a great work. To spend large sums in gathering additional water, and allow at the same time deterioration in the channel that is wholly depended on to conduct the water to the city does not indicate good judgment in the authorities that have it in charge.

On this point I have noticed the complaint of the superintendent of public works as to his difficulty in obtaining from the city authorities even a moderate demand for appropriations for this purpose. Of course the superintendent cannot work without means. And it is certainly strange that while so serious a complaint is made and the danger of failure in a supply of water persists the city authorities should advance large sums to destroy the aqueduct and to make reservoirs for water they cannot obtain if the aqueduct fails.

I will notice one other item in the criticism of Mr. Butler. On page 4 he says, "in the reports of Major Douglass, the *projecting* engineer, and of Mr. Jervis who executed the work" What does Mr. Butler mean by "*projecting* engineer?" If he means that Major Douglass presented the plans of the work, it is strange that in his criticism he notices no plan of Major Douglass' except that in regard to foundations in valleys and so forth; and in this he says the plan of Major Douglass was not adopted. Of course the project in this respect was not followed. If he means projection, as of one who first presented the general project, then he is wrong in ascribing it to Major Douglass. On the 10th of November 1832, the joint committee of the city of New York on fire and water employed DeWitt Clinton, Jr., a civil engineer, to investigate the several sources for a supply of water and recommend such as he considered best for the city. On the 22nd of December 1832, he submitted his report. It was an elaborate presentation of what had been done on this subject. Mr. Clinton does not appear to have made much if any instrumental examination, but confined his attention to such as had been made on the Bronx, etc., and for the Sharon Canal, with others. In these he found a level had been made by Mr. Cartwright of Sing Sing, showing the elevation of Croton River at Pines Bridge to be one-hundred and eighty-three feet above tide. On this, Mr. Clinton very naturally concluded an aqueduct could be carried down the valley of the Croton to the Hudson and then along the valley of the latter to the City of New York. Here Mr. Clinton appears to have first discovered the key that unlocked the project of the Croton Aqueduct. He made the first official announcement and of course was the projector of the enterprise.

The next year [1833] Major Douglass was employed and made instrumental examination, and the next year (1834), Major Douglass and also Dr. John Martineau were employed in further surveys. These engineers made independent reports, and both confirmed the project of the Croton as proposed by Mr. Clinton. This is no disparagement to Major Douglass or Dr. Martineau, who both had the sagacity to follow Mr. Clinton's recommendation.

It may be asked, "How did it happen that two eminent engineers, Benjamin Wright and Canvass White, who were engaged in 1825–26 to obtain water for New York, did not see the value of the Croton?" They were employed by a private company, who did not desire to obtain a larger supply than the market demanded, and hence the Croton appeared too large a project. Also, the city had made rapid growth between 1825 and 1832 and began to consider the matter as one demanded by public authority. This is obvious, and no one acquainted with these engineers will doubt for a moment they saw the Croton project as best calculated to meet the future wants of the city of New York. But that was not what they were called to consider. The Water Company supposed the Bronx River and its tributaries sufficient for their object, and being much less expensive than the Croton, they did not look to it except as an auxiliary for some future wants.

The Water Company did not succeed, and as the city authorities were led to consider the question, the Croton could not fail of a place in their deliberations; and so they engaged Mr. Clinton to make a general examination and recommend such source as he should regard best. This is a brief history of the Croton project. Mr. Clinton had the honor, so far as the documents show, of making the first official announcement of the Croton project. As he was situated, he could hardly, as an engineer, take any other course in view of obtaining a large quantity of water for New York.

More than thirty years had elapsed since the water was introduced into the aqueduct, and I had not looked for further criticism on its engineering. The following reply to a criticism by "M.D." [4] in the *New York Evening Post* seemed to be called for, and I insert it below. (See *New York Evening Post*, November 20, 1874.)

THE CROTON AQUEDUCT

AN HISTORICAL LETTER FROM THE VETERAN ENGINEER, JOHN B. JERVIS

To the Editors of the Evening Post:

In your journal of the 7th of October ult., your correspondent M. D. criticizes your correspondent W. F., in your paper of 15th of September, last, in relation to the Croton Aqueduct. W. F. wrote his article from a hasty gathering of memoranda, and, as very likely to happen in such cases, made some errors, though his statement of facts is, in the main, correct.

If M. D. had confined his criticism to a correction of errors of statement there would be little occasion to reply. But he has thought proper to revive a claim made by the late Major Douglass to credit for the Croton Aqueduct works, which belongs to me as chief engineer of those works.

I have no desire to controvert the claims of Major Douglass or of his friends to credit due him for any work he has executed or plans he may have devised. But it is due to myself, my friends, and to the truth of history that I should state the facts of this case.

Major Douglass and John Martineau, a civil engineer, were employed to make surveys and report on the best method of obtaining a supply of water for the city of New York, and in February, 1835, they made independent reports. (Major Douglass had been employed in 1833.) They both recommended the Croton as the best source. The spring following (1835) the work was authorized by the city. The reports of those engineers were general, mainly relating to the practicality and superiority of the Croton, and discussing, in a very general way, the works that would be required, entering very little into the requisite

4. This was possibly Major Henry Douglass (1825–92), the fourth and youngest son of David Bates Douglass.

details for construction. In this service Major Douglass had spent two seasons. The reports were both able papers of the same kind. The main question was to determine the proper source. I had no responsibility with this question, but as an engineer my attention was turned to the subject as one of general interest, and from observations made in traveling, I had in my own mind decided the Croton was the true source as early as 1825, and so discussed the question with my professional friends. I was therefore prepared to approve the reports referred to. I claim no credit for this, nor do I mention it to detract from their reports.

May 7th, 1835, the work was authorized, and soon after, the Board of Water Commissioners appointed Major Douglass chief engineer of the work and gave instructions to prepare it for contract. He continued chief engineer until the 11th of October, 1836. At this time (the close of his service) it appears there was nothing prepared for contract. Major Douglass had spent most of two seasons, 1833 and 1834, in the preliminary surveys and reports. During this time it was natural to suppose his mind had been engaged on the requisites of such a work, and that he would enter on the duty of chief engineer much advanced in preparation of plans, specifications, and forms of contract. As chief engineer he had in addition the season of 1835 and most of 1836 to get ready for contracts.

Late in September, 1836, I was called on by a committee of the board of commissioners, Messrs. Stephen Allen and Saul Alley, who opened a negotiation with me to accept the position of chief engineer of the Croton Aqueduct. This led to my appointment on the 11th of October following. It is claimed by M. D. that the cause of Major Douglass' removal was "personal and political," and not in any wise from want of confidence in his ability as an engineer. I can only judge of the motives of the commissioners by their acts. No personal communication, either direct or indirect, had previously been held with me, except an answer to a letter of Mr. Allen requesting that I would send copies of specifications, contracts, etc., of the state works— not the least intimation to me in this of any thought that I might be proposed as a successor to Major Douglass. Nor had I a thought of any such proposition. The call on me, as above, I had in no way anticipated. If they were actuated by more personal considerations, it was from no personal relations that I held with them. If their motive was political, as Major Douglass distinctly intimates, they were careful not to place before me, at that or any subsequent time in all my connection with them, any such motive.

If the engineering work had been conducted on political grounds, why did not the Whig board that came into power in 1840 reinstate Major Douglass? At that time there was much adverse criticism of the works, and Major Douglass sent a long communication to the new board, predicting much disaster to follow from the imperfect character of the works, at the same time elaborating the great duties he had discharged and claiming that his plans had not been changed, "except in regard to Harlem Bridge and some minor details," and intimating his readiness to do anything he could to correct the errors that had been made and were in process of execution. This looked very much like asking to be reinstated as chief engineer. Of course, his predictions were based on my unskillful execution of his plans. The criticisms of Major Douglass and others to which I have alluded made a strong impression on the board of commissioners. I well recollect one morning Mr. Samuel Stevens, the chairman of the board, came into the office (his desk and mine were in the same room) with an expression that indicated much anxiety. I was writing at my desk. I laid down my pen to see if I could ascertain the cause. Casual conversation ensued, which soon brought up the aqueduct. Mr. Stevens, with a significant sigh, remarked that it would be sad if, after spending so much money, the aqueduct should be a failure. I replied that it would be sad indeed; that I had no doubt of its success; that my experience and investigation gave me confidence; that it was impossible for me to explain to him, for he could not be expected to follow the scientific reasoning or see the force of experience in such matters; that he must have faith, and if he did not think I was capable of conducting the work successfully it was his duty to engage an engineer on whom the commissioners could rely. Here was a clear case for reinstalling Mr. Douglass if the board had thought proper. I took no measure to influence them other than by a strict attention to my duties as engineer of the works. It is well known the board did not make the change, and the public will judge of the reasons.

The old board, in their negotiations with me, put the question wholly on the ground that they had lost confidence in the engineering ability of Major Douglass to prosecute the work successfully and made no allusion to politics. It is not necessary for me to discuss the reasons on which they predicated their opinions; I merely state that fact as they presented it to me.

M. D. states that the commissioners pulled up Major Douglass' stakes and quarreled with him about the location of the Croton Dam. This appears singular, as the dam is located substantially as Major Douglass proposed it in his report of February, 1835, and I heard nothing of such controversy, or why Major Douglass had so located it in his actual location of the line. The commissioners never showed a disposition to meddle with my stakes.

It is stated by M. D. that the old board did not see the necessity of an engineer department, and refused to him the necessary means to carry forward the work. To me this is strange language, as, upon my entering on the duties of chief engineer, I found such department in full organization, with all the force that appeared to me necessary and, as I subsequently found, more than could be beneficially employed at that time. As the work continued, I found no difficulty in obtaining from the board all the facilities I needed, and they showed no disposition to encroach improperly on the duties of the department. I laid before them from time to time mostly written communications with plans, specifications, and forms of contract which were fully and intelligently discussed and in almost every case were approved. The commissioners did not undertake purely scientific questions, saying they left such to the engineer, who, they supposed, understood them. In all my intercourse with the commissioners, I had no occasion to be dissatisfied with their treatment of my department. I may add that, in all my engineering experience, I have not acted with an authority that was more respectful of the rights and duties of an engineer department than the old Board of Water Commissioners. If M. D. is correct, then this board must have treated him very differently from what they did me. I will add, as due from me, that I found the old board unremitting in their devotion to the interest of the work. In this I must not be understood as slighting the successive boards, with whom my connection was pleasant, and who devoted themselves to the great duties they had in charge.

I now come to the main question of M. D., namely, that Major Douglass devised and prepared the plans on which the work was constructed. It has been noticed that Major Douglass, in his communication to the board in 1840, claimed that the plans of work were his, "except the Harlem Bridge and some minor details." M. D. refers to the preliminary reports of Major Douglass as a confirmation of his assertion that these were the basis of the works. I appeal with confidence to those reports to show that they embrace no "plans of work" as constructed for the Croton Aqueduct.

I will briefly notice the condition of affairs when I entered on the duties of chief engineer.

I have noticed that Major Douglas had been occupied two seasons on preliminary examinations—in addition, the most of two seasons as chief engineer after the work was authorized. I supposed he had in this time made important progress in settling the details and plans of the requisite works. My first acts were to ascertain what he had done. In this I first found that Major Douglass had submitted to the board and they had approved a plan of masonry for the conduit, or waterway of the aqueduct. I inquired if the board considered this plan conclusively settled. They replied it was in my hands, and they would consider such report on it as I should think proper to make. I took up the subject as other duties would permit, and on the 23d of December 1836, I submitted to the board a revised plan of conduct. In August following, I made a further report on this section. In these I did not materially change the inner form or waterway. I so changed the dimensions of masonry as to reduce the expense on this part of the work more than a million dollars. The board accepted my views in these respects, but I had more difficulty in obtaining their approval of my views in relation to the use of quick lime in a portion of the masonry than in any other question, and several interviews on this point had been fruitless in obtaining their approval. Major Douglass had recommended the use of quick lime in part, and some members of the board had been strongly impressed with its propriety. I could not consent to the use of quick lime in any part, and they finally yielded to my judgment. I indulged in no criticism disrespectful of my predecessor. Considering the quantity of water to be provided for and the respective elevations of origin and delivery, I regarded the interior form and size of the conduit as mainly satisfactory. I very well knew that in so long a line of conduit small sectional difference would make a large aggregate in the masonry, and felt the importance of reducing this, so far as consistent with proper stability, and am responsible for the change above noted that has been important in the economy of the work.

The next item I noticed was the location of the line from Croton to Harlem River. This had been made by Major Douglass. One of my first duties was to walk over this line. The elevation of country did not admit to any material extent of two lines, the only exception occuring between Yonkers and Harlem River. The location, for the most part, was a question as to how deep cutting should be taken where the ground admitted a choice. The valleys and ridges that lay across the route, and could not be turned, must be crossed. I made no material change in the line as located to Harlem River by Major Douglass. On this point, I refer to my report of the 31st of December 1836.

The next item presented to my consideration was a specification of work for a tunnel above Sing Sing. I made no written report on this. I said to the commissioners it was so different from my views on that subject that it would be best for me to prepare such as appeared proper to me, and they could judge which should be adopted. My specification and form of contract for all the works, the conduit included, were adopted. No other specification than I notice above was prepared by Mr. Douglass for any of the works, so far as I could find.

In addition to the above, there were several plans by Major Douglass of bridges over ravines. One of these, the most important, was for a valley at Sing Sing. This (as the others) was a plan by several small arches. I prepared a totally different plan, with one elliptical arch of eighty-eight feet span. Not one of these plans left by Major Douglass was adopted.

I certainly felt the responsibility of the work and was desirous of availing myself of any details Major Douglass had worked out, and had not the least aversion to adopt anything he had proposed, so far as my judgment approved. From the long time he had been engaged on the work I did expect to find more progress in preparation. Those noted above were all that I could find in the engineering department, and they were used only to the extent I have stated.

The charge of M. D. that your correspondent W. F. stated that Major Douglass proposed inverted syphons for crossing Harlem River is correct. In this W. F. was in error. Major Douglass, in his report of February, 1835, discussed the question of pipes and bridges for crossing Harlem River and Manhattan Valley, but decided for bridges on the basis of economy. Mr. Martineau proposed pipes as most economical. Neither of those engineers seemed to have had any idea that navigation would be interfered with, and it was on this part that I was severely criticized for proposing pipes. The legislature interfered on the supposition that the navigation would be injured, and required a bridge that should be one hundred feet clear between the surface of the river and underside of arches. It was then my duty to prepare a plan in accordance with law. It is true Major Douglass had proposed a bridge, but nothing could be found in the office in relation to it among the papers of Major Douglass. It was easy to propose a bridge; but to devise a plan and specifications that should provide for all the requirements of such a work, to consider the varied character of the river bed, the depth of water, mud, boulders, sand and rock, the cofferdams and centers of lofty arches, the character and dimensions of masonry, all so provided and executed as to give the requisite stability, was quite a different thing. Whatever it is, the engineering is wholly on my responsibility, and the effort to deprive me of this is not just. It is due to Major Douglass that he did not make the claim as by M. D., but stated that in this respect his plan was not adopted.

I have shown how far the claim of M. D. that the aqueduct was mainly on the plans prepared by Major Douglass is founded in fact. As to their being found in [Major Douglass'] reports, no such are found there. In the most difficult portion of line for location, that on Manhattan Island, or in the preparation of plans for ventilation, waste weirs, culverts, supporting walls, bridges, masonry in tunnels, Croton Dam, receiving and distributing reservoirs, with their influent and effluent appurtenances for the regulation of water, the Harlem Bridge, the pipes at Manhattan Valley, with the specification and form of contracts for these and every other work on the aqueduct, nothing was left by Major Douglass to guide in any respect what I have mentioned. I have no complaint against Major Douglass. He probably left all he had prepared; but it is not right that his friends claim for him devices and plans that he had not prepared.

If the work had not been successful no doubt I should have been held responsible, and the numerous critics (Major Douglass included) would have pointed to their predictions with satisfaction. I have just ground to complain that, while so little of the work followed the designs of Major Douglass, such persistent pretensions should be put forward to deprive me of my just responsibility. I say pretensions, for they are not founded on facts.

In some way, this subject has got into historic form. The preliminary essay to the geological history of New York by the late Governor Seward gives expression to the view claimed by Major Douglass. Appleton's *Encyclopedia* has the same. Mr. King, in his *Memoir of the Croton Aqueduct* (page 143), says: "The plans and details were digested by Major Douglass." Strange delusion, as he might easily have known what a paucity of preparation in these respects had been left by Major Douglass (Mr. King, however differs from Major Douglass in giving me the credit of having well executed the plans). Newspapers have committed the same error. How did these errors obtain currency? Some person must have started them. I cannot suppose those persons referred to designedly misrepresented the facts. No doubt in some way they were misled. A year or two before his death, Governor Seward, on my calling his attention to the errors, stated that he did not pretend to know the facts; that some one, he did not recollect who, had so advised him.

In whatever way any of the persons referred to obtained information they were wholly mistaken as to the facts. In this they have done me great injustice and have been instrumental in perverting the truth of history.

There is one circumstance that must give an impression on this subject, namely, that Major Douglass was first appointed chief engineer under favorable auspices, and during about seventeen months he so lost confidence that the same board removed him. I took the appointment with no acquaintance with the board, and through their administration and three successive changes by political action, with a strong influence for the reinstatement of Major Douglass, yet I was retained until the work was completed, and I retired on my own unsolicited resignation.

The fact that the aqueduct has served its functions more than thirty years shows that my critics were in error, and the several boards justified in trusting to my engineering ability.

After adversely criticizing my works, it is unreasonable that the late Major Douglass should have the credit of my labors after success had been realized.

JOHN B. JERVIS

Rome, N.Y., November 12, 1874

XII

LATER PROJECTS

Boston Aqueduct

The city of Boston had made an unsuccessful effort to procure a supply of water. The difficulty arose from a divided opinion as to which of several projects was the best. A board of commissioners had reported the Cochituate as the best of the several projects, but the opposition of the friends of other projects was so strong that the measure was defeated by a vote of the citizens in the spring of 1845. Not willing to abandon an enterprise so important to the interests of the city, the city council appointed a committee to institute necessary measures to secure a supply of water. This committee decided to select a commission, consisting of one person from the city of New York and one from Philadelphia, to make an investigation of all the projects that had been presented, and report such as they should find to be the best.

For this, they engaged Mr. Walter Johnson of Philadelphia (a professor in a scientific institution) and the writer of this Memoir. I went to Boston, and there met Mr. Johnson and the members of the water committee.

After making some general examination, it was apparent to me that the duty was essentially a matter of engineering, and that under the circumstances existing, it would demand very careful examination and great prudence to bring out successfully the varied features of the proposition. Not having any previous acquaintance with my associate but the fact that he was not professionally an engineer, I concluded not to enter upon the service unless the engineering was placed entirely in my hands. This decision was not favorably received by the committee of Mr. Johnson, but I thought "one poor general was better in command than two good ones," so I adhered. The committee urged that it wanted the moral force of two commissioners. "Very well," I said, "the report may be made in the plural, and I shall have no objection to Mr. Johnson signing it with me, and he may be charged with certain duties that will not conflict with the general engineering." To this the committee assented, and the investigation proceeded.

I engaged Mr. Henry Tracy, a civil engineer of experience and ability, to take charge of the surveys and estimates, a service he performed with prudence and ability. In about five months, the examinations and report were completed by the recommendation of the Cochituate as the best source.

As a suggestion that may be useful to the younger members of the profession—I kept strictly my own opinion until it was tested by the result of a full examination; not even my confidential assistant ever obtained from me the least expression of my views. This was not because I was afraid to trust him, but that he might say at all times he did not know my opinion. In such a case, there is great anxiety with the different parties to obtain knowledge of the prospect of their respective projects, and the least intimation would inevitably lead to discussion. At one time, the committee was impatient to get my views. My associate had been rather leaky, but he was unable to quote me, and the committee appeared to think I should at least give them an intimation. I replied they would know my opinion when they had it in writing at the end of a report. They were not satisfied, and seemed to think I had not treated them with proper deference. In the final result, they expressed themselves as well satisfied with the course I had pursued. The report of the commissioners was generally satisfactory, and the work was authorized to be constructed.

The work for the aqueduct was commenced in the spring of 1826. The Board of Water Commissioners was composed of three members, Messrs. Nathan Hale, James Baldwin, and T. B. Curtiss; Mr. E. S. Chesbrough,[1] was appointed engineer of the

1. Nathan Hale, editor, businessman, and state politician, was, at the time, president of the Boston and Lowell Railroad (contructed by J. F. Baldwin) and editor of the *Monthly Chronicle*. James F. Baldwin, son of the famous Colonel Loammi Baldwin, builder of the Middlesex Canal, was brother of Loammi Baldwin II, a great pioneer civil engineer who in 1835 completed his famous report on introducing pure water into Boston. E. S. Chesbrough, a lead-

conduit and its appurtenances, and Mr. William E. Whitwell, of
the piping in the city. The writer was engaged as consulting
engineer, and continued to the close of 1848 when the works
were completed and the water introduced in the city with en-
thusiastic demonstrations by the citizens. Any further description I properly leave for the acting engineers who had charge
of the construction of the work.[2]

THE HUDSON RIVER RAILWAY

Some surveys and estimates had been made for this enterprise,
mostly by the aid and influence of Mr. James Boorman,[3] a wealthy
and spirited merchant of New York. The project, however, did
not obtain much favor with the public. It was generally regarded
as unpromising in view of the competition of the excellent navigation of the Hudson River.[4] In the summer of 1845, Mr. Boorman called on me to enlist my services in the examination of the
project. I had a favorable opinion of the enterprise, but well understood the difficulty in convincing capitalists that it would be
a commercial success. The general character of the country for
such a work clearly indicated that it would be expensive, and the
competition it must meet from a steamboat navigation conducted
on a magnificent scale led the public to regard it as an extremely
hazardous, if not a hopeless enterprise. After considering the
proposition of Mr. Boorman, I decided to engage upon the work,
and during the ensuing autumn employed Mr. Henry Tracy to
make such surveys as the season and limited funds would permit.

Having given my attention to the project, I devoted much

---

ing American sanitary engineer, began his career as a chainman on the first
surveys of the B&O Railroad and, after an early career in railroads, received the
assignment Jervis mentions here—engineer of western division of the Boston
Waterworks.

2. Jervis selected the preceding extract from his *Memoir*.

3. James Boorman was an English-born merchant, banker, and railroad pro-
moter who became a leading philanthropist in New York City.

4. The Hudson River traffic (New York City–Albany) had been monopolized
by the Robert Livingston–Robert Fulton partnership for many years after the
1807 voyage of the "Clermont." By the time the Hudson River Railway was
seriously proposed, the monopoly had been broken, and highly competitive
boat lines were providing fast (ten to twenty miles per hour) and reliable
passenger and freight service.

time to gathering and analyzing statistics, mostly as to passenger traffic. Mr. Tracy conducted the surveys with great skill, and for the limited means he had, obtained important information. His services were directed to the section between New York and Fishkill, about sixty miles. For the balance of the line, the survey of Mr. Morgan—which left the river at Fishkill, and pursued an inland route—and his report was the basis of the report I made to a company of citizens in January, 1846.

Under this report, a charter for a company was obtained and the corporation organized. Generally the public mind was skeptical, and subscriptions to the capital stock progressed slowly. The indomitable energy of Mr. Boorman succeeded in raising the amount of three million dollars. Loans were depended on for the balance required.

From the general want of confidence in the undertaking, the organization was not perfected until late the following winter. In the spring of 1847, the board of directors appointed the writer chief engineer of the railway, and proceedings were had to commence the work soon after.

As surveys for actual location proceeded, it was soon evident there were obstacles to be encountered that had been very imperfectly developed by the limited character of the original surveys. The rock cuttings in the Highland districts proved to be unusually difficult of excavation. Very little earth could be had in this district for the fillings, and to make these of rock cuttings, especially in those places where the fillings sunk to a great depth in the mud, was very expensive. The long line of exposure to the surf of the river was a serious matter. The great difficulties of the work lay between New York and Poughkeepsie. Above the latter place, the rock was of a slaty character and easily excavated, and the shore had less exposure to the action of the river.

After a more careful examination of the inland route, and also by a survey of the river route made by Mr. John T. Clark, an assistant engineer, I made a report to the board of directors in January, 1848. In this, I recommended the adoption of the river route for the whole line. This report was approved, and the work has been constructed on the river route. It was open for transportation to Poughkeepsie about the close of 1849. For the most part, this section of the railway was graded for a double track, but only a single track was laid at this time. No work of construction had been done above Poughkeepsie.

In August, 1849, I resigned the position of chief engineer, and Mr. W. C. Young [5] was appointed in my place. I continued, however, as consulting engineer until the spring of 1850. At this time, I found that my views in regard to the management and operation of the railway did not harmonize with those of Mr. Boorman, and I therefore resigned and have had no connection with it since then.

It is impossible at this day for any one not engaged in the persistent efforts of the time to appreciate the hardy character of this enterprise. The financial and engineering difficulties were very great. In regard to the former, I can hardly conceive that more could have been done than was effected by Mr. Boorman and a very respectable board of directors. In regard to the latter, the natural impediments were rendered especially severe by difficulties of finance. These in the face of competition from steamboats of great magnificence effecting very cheap transportation, and the incredulity of the public mind as to the capacity of the railway to maintain such competition, rendered the prosecution of the work extremely embarrassing.

The landed proprietors along the route were in general very hostile to the enterprise, declaring it would destroy the natural beauty of the country as well as fail in its commercial object. On this point I had claimed that the natural scenery would be improved; the shores washed by the river would be protected by the walls of the railway; and the trees, no longer undermined and thrown down by the river surf, would grow more beautiful; and that the railway thus combining works of art with those of nature would improve the scenery. For this I was freely reproached as scarcely less than a barbarian. As to my very modest statement, that the railway trains would be able to travel at the

5. William Clark Young graduated in 1822 from West Point but left the army in 1826 to practice civil engineering. He had been chief engineer of the Utica and Schenectady Railroad in 1833.

rate of thirty miles per hour, I was called to account in some very decided poetic effusions.

I wrote an article—published in *Hunt's Merchants' Magazine*, in November, 1846—in which, discussing the subject generally, I took the ground that not only along the Hudson, but also on other steamboat routes, the railway would be a successful competitor. In this article, speaking of railways I said, "The system is viewed as one that mocks the age, its progress has startled the most cautious, its developments are revolutionizing the social and commercial affairs of mankind." This was criticized as a very presumptuous statement. I notice these matters as curiosities of the time. At this day it would be difficult to find any landholder, or steamboat man, to assert the same things. The railway has traveled on and more than reached the position then claimed.

European Tour

On the closing of my connection with the Hudson River Railway in the spring of 1850, I concluded to make a trip to Europe. My health had been impaired by severe and long continued labor, and I had for some time indulged a desire to make such a tour; hence I embraced the opportunity. The trip was of great interest, and afforded me much pleasure.

It so happened that while I was there, one of the large tubes of the bridge over the Menai Straits was to be launched.[6] Mr. Robert Stephenson kindly gave me an invitation to witness the operation. The spectacle was highly interesting in itself and was followed by an invitation to dine with a party of English engineers, an occasion I enjoyed very much. I noticed that they far excelled me in making speeches, which was quite the order at

6. Connecting the west coast of England with the Island of Angelsy, the Britannia Railroad bridge was the marvel of its time. Its two main spans of four hundred sixty feet were supported by towers almost two hundred feet above the water. The huge wrought iron tubes had been the product of the best British engineers, Stephenson, Fairbairn, and Hodgkinson. Begun in 1845, the last tubes (each main tube weighing sixteen hundred tons) were being raised into final position from pontoons during the summer of 1850 when it was finally completed for two-way traffic.

dinner. It was an art I had not cultivated, and I expressed to them my surprise at the skill they exhibited in this respect. It is not worth while to go into particulars of this tour. I will simply state that my observations were mostly on engineering work which interested me very much.

I returned [in August] after an absence of four months, greatly improved in health, and immediately engaged in the construction of the Michigan Southern and Northern Indiana Railways.

MICHIGAN SOUTHERN AND NORTHERN INDIANA RAILWAYS

They were practically one railway, about two hundred and forty-six miles long, extending from the head of Lake Erie to Chicago on Lake Michigan; about sixty-six miles had been constructed by the State of Michigan, with a timber rail and iron plate. It was now in the hand of an incorporated company which had a section of eighteen miles nearly completed, and laid with iron rails.[7] I spent most of the winter of 1850–51 on the line of this railway. There remained about one hundred and sixty miles to be constructed. The route was very favorable. This portion of the work was entered upon in the spring of 1851 with vigor, and in about one year, a line was opened to Chicago. The principal difficulty was one of finance. Though the work was of very moderate cost, eastern capitalists were very reluctant to enter on an enterprise in the western states [those between the Ohio and Mississippi Rivers] as several had repudiated their debts and this impaired confidence in the honesty of their institutions.

During the summer of 1851, I was engaged as president in the construction of the Chicago and Rock Island Railway, extending (as a continuation of the Northern Indiana Railway)

7. It should be noted that there was not a rolling mill in America that could produce an iron rail until the summer of 1844, at Mount Savage, Allegheny County, Maryland, and these rails were of poor quality, compared with the imported British rail. Reliable steel rails were not commercially available until after the Civil War. The constant maintenance of the timber-iron bar rail gave a tremendous impetus to the development of an all-iron (and later steel) rail during the 1840's.

from Chicago to the Mississippi River at Davenport, one hundred and eighty miles. This was also an easy line to construct, and was brought into use in 1854.

I continued my connection with the Michigan Southern and Northern Indiana Railway until the spring of 1858, when it became obvious my views of management could not prevail. With the exception of incidental services, I had no professional engagement until December, 1861.

PITTSBURGH, FORT WAYNE AND CHICAGO RAILWAY

This road was four hundred and sixty-eight miles in length, extending from Pittsburgh to Chicago. It had been constructed under financial embarrassments, and as usual in such cases, was poorly built. There were first and second mortgages on the railway for about $10,000,000 and a floating debt of between $2,000,000 and $3,000,000; the capital stock was about $6,000,000 —making a total liability of about $18,000,000. The company defaulted on their interest, and after two or three years, the bond holders foreclosed and sold the railway in the early part of the autumn of 1861. From inferior construction and inadequate maintenance, the line at the time it was taken in charge by the trustees of the bond holders was in very poor condition.

In this state of affairs, I was appointed general superintendent [December, 1861]. The stock and floating debts were very honorably provided for by the trustees. For the floating debt, a third mortgage was given to the holders of claims, conditioned that no interest should be paid until a sufficient surplus had been earned therefor, after providing interest and sinking fund for the first and second mortgage bonds. After the net income should thus provide as above for all the bonded debt, then the stockholders should be entitled to any balance obtained from net income. I mention these particulars as very creditable to the trustees, and in striking contract with the action of bondholders in some other cases.

The capital stock was very low in the market. I was informed that it had been sold at eight percent. Not long after I had entered on the management, a stockholder enquired of me to know if I thought advisable to sell at twenty percent. I replied that I could not advise in such cases—that I regarded the prospect of traffic as good, but the poor condition would require large expenditures to put the railway in even reasonable order.

The trustees were extremely solicitous to secure such net earnings as would provide for the interest and sinking funds of the first and second mortgage bonds, saying that beyond such provision they were willing all earnings should go to the improvement of the railway. In order to secure their object, the trustees required a monthly sum to be placed in their hands.

I will not detain you with further details, but merely give the financial result. The first year passed with decided expression of satisfaction, and at the end of the second year, there were further net proceeds. After two and one quarter years, I resigned the superintendency, but continued to act as engineer until 1866. The bonds in interest and sinking funds had all been provided for in these two and one quarter years, and such surplus was in hand that soon after, the board of directors declared a dividend on the stock at the rate of ten percent per annum, which rate has not since been reduced.

XIII

PROFESSIONALISM AND THE ENGINEER

THE OPINION is sometimes advanced that men cannot be induced to work for the public or for a corporation as faithfully as they do for individuals. This depends on the supervision they have. If the foreman in charge of the work regards it proper to keep a chair to sit in and under the shade of a tree, or if the sun induces him to carry an umbrella over his head, his men will no doubt regard it proper to avail themselves so far as they can of similar indulgences and do as little work as may be practicable to secure enrollment on the payroll. This is no extreme case, for I have often seen it in work done for the City of New York. Civil service did not then require the abandonment of sound business principles of economy. Commissioners were appointed from the party in power; but when they entered on their duty, they employed men who were supposed best qualified to do the work required without regard to their politics. So engineers, contractors, and other agents were selected without regard to their political faith.

There is still another class of foremen who are men of character; who stand round and give directions with so much superciliousness that they call a man out of his work to do the least little thing which they could easily do themselves, as lending a hand for an occasion to aid the men in a lift. But to handle a bar or a lever they regard as inconsistent with their dignity and so break in on some regular party, deranging their work to save the foreman's views of dignity. I have often seen this done, when the foreman by a moderate exertion of his own powers would have spared the derangement of the work of a gang of men near by. This class will stand over their men with folded hands, giving directions but careful that they be not caught in doing any useful thing by manual effort, regarding such action as a thing beneath their position.

When merely political and sometimes corporate influence controls, the classes above noted are very likely to be placed in charge; the political from notions of patronage, and the corporate by influences that sometimes introduce relatives, and other friends without respect to the economy of the work.

In the supervision of a party of workmen, the foreman should look well to the order of his men, and see that each has his proper position and is doing his duty in his place. It will often happen that some incidental thing must be done to keep the party in regular work, and can easily be done by the foreman who has a prompt eye to his work. In such case he should not break in on any gang of his men to do such incidental work. A good foreman will look well to the arrangement of his men to see that all move harmoniously to accomplish their work. It should not be expected the foreman will place himself in the position of a regular workman unless it be with a small party in which he may, for the most part, work with his men. His first purpose should be to see that his men are well arranged for this work, and then do whatever he may find to do, and so carry forward his work as to accomplish the most economical result. For a man to stand with folded hands as a foreman over a small number of men always appeared to me a very trying position requiring a good constitution to endure it. This notion of dignity is often manifested by teamsters, who refuse to use the shovel while their load is being filled. In fact it is often seen in other workmen who show their sense of dignity by refusing to do something not regarded as belonging to their work. It is well that each workman has his own work, but it often happens that some exigency demands a modification, and and when this occurs the progress of the work should not be embarrassed by any peculiar notions of dignity, either on the part of the foreman or the workmen under his charge.

A good supervisor of work must feel responsible for its judicious management and bring his mind to feel the same interest in its success as he would if doing for himself. This is demanding merely an honest purpose which no upright man will refuse. This is a cardinal requisite for a man who undertakes the responsibility of conducting the business of others. Such men enjoy the success of their operations as highly as if the result were their own.

Next to fidelity to his trust the supervisor of work for others should have the industry and experience required for the special charge he undertakes. The economy that will result will be the same whether he serves an individual, a corporation, or the government. With a proper supervision the workmen will be equally industrious in either case. The supervisor, if he has any tact, will easily inspire the workmen with his own spirit, and from such experience I am fully of the opinion that the economy of work will be secured by the ability and fidelity of the supervision, irrespective of the question as for whom the work may be.

The real point in the matter is when an individual proceeds with work, he is careful to select the most reliable man he can find to conduct it. The same is true with government and corporation work, when the authority for its management is in the hands of capable and honest men. Officials of this character have found no difficulty in securing the most economical results. I do not by any means claim this is always secured; but failure does not come from the fact that there is any inherent difficulty in the case, but from want of proper characteristics in the officials who are entrusted with the charge. In the work placed under my several charges as I have described, I had the pleasure of acting under officials who looked steadily to the interest committed to their care. No political partisan presumed to interfere in the supply of services or material. The only action that came under my observation was to secure the best that could be obtained for the work to be done. Under such circumstances, there is no difficulty in prosecuting government or corporation work with the same economy and efficiency as can be done by individuals for themselves. It is the want of these requisites above mentioned in the officials in charge that makes government and corporate work expensive. The workmen will do well if they have proper supervision.

ENGINEERS AS RAILWAY SUPERINTENDENTS

Now, it may be said, why all this about the superintendency of
a railway—engineers are not much engaged except in construction? My object is to show that an engineering education fits
a man best for the superintendence [management] no less than
for the construction of a railway. I regard it an error that the profession has not given more attention to superintendence of operations. It demands the same thoughtful analyzing mind to operate
what it constructs. I have presented this case to impress the importance of employing the best engineering experience and ability in the superintendence of railways. The case is a plain one.
Say, management has failed to pay interest, and the railway was
sold by foreclosure of mortgages. An engineer takes charge, and
in two and one quarter years this is all changed and the property
becomes a success. But why an engineer? A true engineer, first of
all, considers his duties as a trust and directs his whole energies
to discharge the trust with all the solemnity of a judge on the
bench. He is so immersed in his profession that he has no occasion to seek other sources of amusements, and is therefore always
at his post. He has no ambition to be rich, and therefore eschews
all commissions that blind the eyes and impair fidelity to his
trust. His ambition is to make his trust a success to the proprietors. His education fits him to analyze facts, and more successfully than a layman to produce the best results of administration.
To reach the best results of railways, they must be put under the
superintendence of competent engineers.

It may be inquired, why should not engineers seek to be rich?
Certainly, they need the benefit of pecuniary resources as well
as others, which is not denied. "Rich" is a relative term. The
man who, by a proper economy, is able annually to lay something
by as an investment from his earnings is only exercising the prudence necessary to provide for age or infirmity. This all men
should do, and it is especially important for an engineer that by
this means his mind may be at ease in regard to the future. What
is meant by a thirst to be rich is a speculating turn of mind which

is sure to prevent any special eminence in an engineer. He should be careful to secure a proper salary and then give his undivided attention to the duty he assumes. I have known several engineers who were always on the alert for speculation, but I have not known one such to secure any valuable standing as an engineer. To this it may be said engineers have not, in all cases, succeeded in conducting the operating business of railways. The same may be said of lawyers, or any other profession, and it is just as important that the party in interest select a competent and faithful engineer as a competent lawyer. The same discretion is necessary in one case as in the other.

Being now past four score years of age, I can have no other motive than the public good in the above remarks. I make them in view of much personal experience, which has brought me to the conclusion that the railway and kindred public works will not realize their legitimate benefits until their supervision is committed to competent engineers. In order for this, it is necessary to cultivate the profession so as to secure the qualifications and character before named, to carefully avoid all theories that are not based on well-established and thoroughly analyzed facts, and to be very cautious before applying any theory, to see that all the facts involved are obtained, and that none are hastily and inconsiderately assumed. The profession is one of great importance to the public interest, and no pains should be spared to fully qualify those who pursue it for every engineering emergency and to fit them for the discharge of every duty with scrupulous fidelity.

CHRONOLOGY OF JERVIS'
PROFESSIONAL LIFE

1795—Born December 14, at Huntington, Long Island, New York.

1798—Moved to Fort Stanwix, New York.

1817—Hired by Benjamin Wright in the survey party of Nathan S. Roberts as an axeman to work on the Erie Canal.

1818—Promoted to rodman with Roberts; joined survey party of David S. Bates; met Canvass White; worked as a stone-weigher.

1819—Made resident engineer; middle section of canal completed from Utica to Rome; promoted to surveyor with Bates.

1823—Promoted to superintendent of operating division (fifty miles long).

1825—Appointed principal assistant under Wright for the Delaware and Hudson canal and railroad system; surveyed canal route with John B. Mills.

1827—Appointed Chief Engineer, Delaware and Hudson; surveyed and located gravity railroad from the canal terminus at Honesdale, Pennsylvania, to Carbondale, Pennsylvania.

1828—Letter to Horatio Allen with specifications for locomotive procurement.

1829—Arrival of first locomotive in western hemisphere from England; Honesdale trial run with "Stourbridge Lion," Horatio Allen at throttle.

1830—Appointed chief engineer of the Mohawk and Hudson Railroad and later of the Schenectady and Saratoga Railroad.

1831—Ground broken for the Schenectady and Saratoga Railroad.

1832—West Point Foundry built the "Experiment" locomotive according to Jervis' plans, employing his swiveling, four-wheel, bogie truck to support the forward end of the locomotive achieving a speed of eighty miles per hour.

1833—Appointed chief engineer of the Chenango (New York) Canal. Employed artificial reservoirs for water at summit level; first bogie truck locomotive shipped from Robert Stephenson and Company, England—the "Davy Crockett," used on the Schenectady and Saratoga Railway.

1836—Appointed chief engineer of the eastern division of the Erie Canal enlargement; accepted position as chief engineer for the Croton Water Supply System, which included the Croton Dam, the Ossining Aqueduct, the Harlem River Aqueduct, and the distributing reservoir at 42nd street and 5th Avenue in New York City.

1840—Attended convention in Philadelphia regarding the formation of the Institute of American Civil Engineers.

1842—Published *Description of the Croton Aqueduct.*

1846—Appointed consulting engineer for the Cochituate water works (Boston); *Report on the Hudson River Railroad.*

1847—Appointed chief engineer for the Hudson River Railroad to Poughkeepsie.

1848—Published *Hudson River Railroad* and *Report on Location.*

1850—Became chief engineer for the Michigan Southern and Northern Indiana Railroad during which time he also constructed the Chicago and Rock Island Railroad; published *Report . . . Michigan Southern and Northern Indiana Railroad.*

1854—Became President of the Chicago and Rock Island Railroad.

1855—Defeated as the Democratic candidate for New York State Engineer.

1856—Published *Letters Addressed to the Friends of Freedom and the Union.*

1857—Published *Report of . . . Railroad Bridge over the Mississippi River at Rock Island.*

1861—Became general superintendent of the Pittsburgh, Fort Wayne and Chicago Railroad until 1864 (consultant until 1866); published *Railway Property: A Treatise on the Construction and Management of Railways.*

1868—Organized Merchant Iron Mill, Rome, New York; published *Letters for the Roman Citizen . . . Currency and Debt of the United States;* elected an Honorary Member of the American Society of Civil Engineers.

1877—Published *The Question of Labor and Capital.*

1878—Received LL.D. from Hamilton College, New York.

1885—Died, January 12, in Rome, New York, at the age of eighty-nine.

INDEX